STRANGE INTRUDERS

The thing is that when I bit her . . . she didn't scream, she didn't cry, she didn't react at all in pain. . . . It was as if I took a bite out of a plastic dummy or mannequin that was made of rubber or something.

The expression on her face was like . . . this isn't the way it's supposed to be . . . this isn't the way it's supposed to happen. You've done this wrong. . . .

You know when you look at someone and you see a sparkle in their eye, you know they're alive . . . I didn't see that . . . there was something missing.

[But] they knew what they were doing. They were there for a reason. . . .

HAIR OF THE ALIEN
ALSO AVAILABLE AS AN EBOOK

HAIR

OF THE

ALIEN

DNA and Other
Forensic Evidence
for Alien Abductions

BILL CHALKER

PARAVIEW POCKET BOOKS

New York   London   Toronto   Sydney

PARAVIEW
191 Seventh Avenue, New York, NY 10011

POCKET BOOKS, a division of Simon & Schuster, Inc.
1230 Avenue of the Americas, New York, NY 10020

Library of Congress Cataloging-in-Publication Data
Chalker, Bill.
    Hair of the alien : DNA and other forensic evidence for alien abductions
/ Bill Chalker.—1st Paraview Pocket Books trade pbk. ed.
        p. cm.
    Interview with Peter Khoury
    Includes bibliographical references and index
    ISBN-13: 978-0-7434-9286-7
    ISBN-10:   0-7434-9286-2
        1. Human-alien encounters—Australia—Case studies. 2. Alien abduc-
tion—Australia—Case studies. 3. Evidence, Expert—Australia—Case
studies. 4. Forensic genetics—Australia—Case studies. I. Khoury, Peter. II.
Title.

BF2050.C46 2005
001.942—dc22   2005047588

First Paraview Pocket Books trade paperback edition July 2005

10   9   8   7   6   5   4   3   2   1

For information regarding special discounts for bulk purchases, please
contact Simon & Schuster Special Sales at 1-800-456-6798 or
business@simonandschuster.com.

To all those who have been touched by the alien abduction phenomenon, I hope this work will help anchor your journey in reality, whatever that reality turns out to be.

Entering "The Land of Illusion":

Truth becomes fiction when the fiction's true;
Real becomes not-real when the unreal's real.

Cao Xueqin (c. 1760)
The Dream of the Red Chamber
from the translation by David Hawkes
published by Penguin Classics, 1973, from
The Story of the Stone: Volume 1, The Golden Days

CONTENTS

CONTENTS

INTRODUCTION

The abduction story that Peter Khoury tells is awash with many of the bizarre features typical of alien abduction stories from around the world. But his case is spectacularly unique in at least one extraordinary sense. It features a piece of evidence unprecedented in the annals of UFO research and investigation: a very unusual biological sample, a fragment of strange hair to be exact, which in a remarkable but paradoxical way confirmed the truly controversial and provocative nature of the episode that led to its recovery. This hair sample became the focus of the world's first DNA investigation of biological evidence implicated in an alien abduction.

I will describe this extraordinary case that has provided us with fascinating genetic evidence, which forms the nexus of a remarkable breakthrough and the cornerstone of this strange odyssey. With the help of biochemists, I have carried out a thorough forensic investigation of this case, utilizing DNA profiling, coupled with an archaeological-style DNA investigation. The results that have emerged from our work are nothing less than startling.

Hair of the alien? This is no bizarre variant of a "hangover cure" for UFO obsessions. But it may provide a "cure" for the lack of physical evidence that has plagued alien abduction research for the past four decades. What we are dealing with here is a fundamental refocusing of our perceptions of alien reality.

There is no doubt that the small bald gray alien is a powerful icon that represents our society's image of an alien lifeform. What is not widely understood, even by many knowledgeable people in the UFO field, is that the dominant image of the bald gray alien is to a great measure an artifact of the popularization of the "gray" through the efforts of some key researchers, the media, and in particular one prominent abductee—a term given to an individual who claims to have had an alien abduction experience. So pervasive has been the impact of this popularization that the validity of a claim of contact with aliens is now often judged by how closely it conforms to the gray alien stereotype. But like most stereotyping efforts, the media-driven march of "gray" conformity is far from the truth and to some extent has misdirected a generation and perhaps delayed a more realistic understanding of the situation.

With its striking image of an alien face on the cover, *Communion*, Whitley Strieber's 1987 account of his claimed encounters with aliens, sold millions of copies and solidified in public consciousness the perception of the alien as bald with large dark eyes. Thousands of accounts describing aliens emerged worldwide in response to this extraordinary public profiling of the alien image. But Strieber had suppressed one key detail of his alien encounter to see if it would be validated independently in subsequent witness descriptions of alien beings. That detail was the presence of hair in descriptions of aliens. According to Whitley Strieber, only one case emerged in this human tide of encounters with aliens that strikingly matched his own experiences with aliens, sharing the key element of the presence of hair. That case turned out to be from Australia. This largely hidden aspect of the alien image may be a potent argument for a sociological dismissal of the whole alien saga, on the grounds that if thousands didn't report it, then the majority may have been delusional or just jumping on the alien bandwagon. Such an argument is found wanting, however, when this suppressed aspect emerges as the central piece of evidence from Peter Khoury's encounter, namely the found "alien hair" that was subjected to DNA testing.

Our current understanding of the alien image has been profoundly skewed. The gray alien icon is clearly not the full picture. Part of that complex and bizarre picture includes beings that are far closer to us in appearance, featuring unambiguously hirsute elements, than we have ever imagined. What emerges in this investigation is a picture of a phenomenon that redefines our origins and our place in the universe. This investigation goes further than any other before it, because for the first time the fantastic has been anchored in a strange and disturbing reality. After decades of neglect, science is now colliding head-on with the alien abduction controversy.

Whatever is going on, one thing seems clear. These experiences seem to be a form of intrusion—an intrusion on our cozy grasp of prosaic reality—and if the experiences are real events, the intrusions are a physical assault on the individuals who experience them. The assumption is that if such experiences are more than imagination, there must be tangible evidence. Knowing what to look for and utilizing the appropriate tools to reveal the nature and implications of that evidence is the key to the problem. Ultimately the application of science, mediated with a forensic focus, will hold the key to determining the reality of alien abduction claims.

Our hunt for clues and answers to this alien mystery features the DNA profiling techniques used by law enforcement organizations such as the FBI and increasingly by police around the world. Such work was made possible only through a remarkable breakthrough, the PCR (polymerase chain reaction) procedure. Dr. Kary Mullis is credited with the initial concept that led to the PCR technique, a discovery that ultimately earned him the 1993 Nobel Prize in chemistry. The technique rapidly revolutionized biochemistry and opened the way for extraordinary new opportunities for understanding ourselves as a species.

Dr. Henry C. Lee, one of the leading forensic scientists in the United States, and Frank Tirnady, describe "the new DNA paradigm" in their book *Blood Evidence: How DNA Is Revolutionizing the Way We Solve Crimes*. Lee and Tirnady end a long list of striking applications of DNA evidence by noting its ability to resolve "contentious public debates" and illuminate "countless other controversies involving biological issues,"[1] both of which are relevant to the controversial issues in this book. The striking utility of DNA analysis rests on one principal fact. "It permits the differentiation of one person from another on a genetic level," Lee and Tirnady explain. "It can also distinguish one plant from another, one animal from another, and even one viral or bacterial strain from another. DNA-identification is predicated upon the genetic uniqueness of each individual (excepting identical twins, triplets, or clones), a uniqueness driven by the engine of sex."[2]

"Forensic DNA analysis is not a single technique," Lee and Tirnady point out. "It is a swarm of techniques that have been coaxed out of the vast expanse of the human genome. . . . [Most] of the forensic techniques are generally powered by polymerase chain reaction, or PCR, a revolutionary process that permits the faithful reproduction of small, sometimes even vanishingly small, amounts of DNA. The process produces large amounts (identical copies) of DNA . . . [and] allows investigators to obtain a DNA profile from postage stamps, coffee cups, and even single hairs."[3,4] Then, quoting the U.S. National Institute of Justice review of DNA evidence, they note: "[DNA] has passed the test. The scientific foundations of DNA are solid. Any weaknesses are not at the technical level, but are in possible human errors, breaks in the chain of evidence, and laboratory failures."[5]

The applications of PCR and the DNA analyses to the alien abduction phenomenon are really about a potential breakthrough. Certainty will come with time and any possible replications, but for the moment I would be very pleased if this work helps illuminate this extraordinary controversy with the promise that science can light the way through the dark morass that currently swamps the alien abduction debate.[6] Lee and Tirnady conclude, "The singular contribution of forensic DNA analysis is that it has made everything biological a potential clue, and this adds an entirely new dimension to figuring stuff out."[7] While DNA techniques are indeed an impressive breakthrough, the whole forensic scientific approach will go a long way to providing understanding and certainty in this hotly debated field.

The PCR concept is not the only indirect contribution

that controversial scientist Kary Mullis makes to the saga told in this book. Mullis has a further unique claim to fame. He is the only Nobel Prize laureate to admit to a possible alien abduction experience. His bizarre experience, described in this book, is based on his descriptions and my discussion with him.[8]

One of the great pioneers of scientific detection, or forensic science, was Dr. Edmond Locard. He is perhaps best remembered for the basic principle behind all of forensic science: *Every contact leaves a trace.* While Locard clearly had in mind crime committed by human felons, the application of the principle in the very controversial area of "alien abductions" is of critical significance.[9]

If alien abductions really occur in the manner described by thousands of people around the world, then at the very least a "crime" against the sovereignty of humankind is being perpetrated. Most of the claims of alien abduction describe situations that are clearly against the will of the individuals involved. Thus the "crime" interpretation and the forensic approach it inspires are well founded. But beyond the striking biological perspective is a much broader and complex dimension to the forensic approach. As in classic crime scene investigations, the evidence for the reality of abduction claims can come from diverse avenues, such as the biological, physical, intuitive, and just plain commonsense. There is also a cautionary element to forensic investigations; they require considerable care. Even while looking for traces of alien "contact," we must carefully examine the impact of the "contact" that comes with the intrusions of researchers, investigators, and other participants into these bizarre situations. Their involvement can often give a less than satisfactory or reliable dimension to the more prosaic senses of "contact" with possible alien realities. Not every contact leaves a *reliable* trace. Every aspect needs to be carefully considered. Beyond these cautionary elements, the idea of a "contact" resonates powerfully with the otherworldly implications of alien abduction experiences.

Here, then, is a factual account of a series of strange events that suggests a convergence of *C.S.I.: Crime Scene Investigation*[10] and *The X-Files*.[11] But this is not fiction. It represents a powerful case for the forensic and scientific approach to the alien abduction experience. At its heart is an extraordinary event steeped in the bizarre and the unusual, an investigation anchored in a forensic scientific perspective, and an odyssey that will yield some remarkable speculations and new viewpoints on what clearly is one of the great mysteries of our time.

CHAPTER ONE

Shattered World

PETER KHOURY'S LIFE UNDERWENT A PROFOUND CHANGE ON July 12, 1988. His world was shattered by the intrusion of something he simply could not comprehend.

I couldn't scream. My mind was functioning, but I couldn't move. This is evil. I'm paralyzed. I thought, is this the devil? Nothing like this had happened to me before. This paralysis, it was physical. I felt it crawling up my body, a rush of pins and needles. I was petrified. I was overwhelmed by the thought that I would never walk again. Then I became aware that I was not alone. There were three or four ugly figures only about three to four feet tall near me. Their faces were very wrinkled and shiny dark black in color. My fear, already overwhelming, was now soaring out of control.

Peter Khoury experienced a classic alien abduction, but he is not the typical alien abductee. Born in 1964, in Lebanon, he migrated with his family to Australia in 1973. He met his future wife, Vivian, while at school in 1981. They married in 1990 and now have two children, Stephen (born in 1995) and Georgia (1998). Peter works in the building trade, where he has had his own business in cement rendering. Peter's Lebanese relations and his wife's Greek relations form a large and close family unit.

Peter is one of the legions of the abducted. A growing number of people from all walks of life, all around the world, have been confronted by the experience of being taken, over-whelmed, or assaulted by something that is intensely strange. It seems to them that they may have experienced the unbe-lievable and the preposterous, because the perpetrators are seemingly not your usual human beings. Some appear human, but the circumstances in which they appear are dis-tinctly alien. Others are totally alien in both countenance and circumstances.

For most people, the alien abduction phenomenon is a ridiculous subject. It is inextricably mixed up with something else many regard as science fiction, namely the UFO mystery. But some researchers suggest that the alien abduction is actu-ally a hidden epidemic. There may be a very large number of people who have had this experience. Many of them have come forward, reporting this disquieting experience. Most seek understanding and support. Few receive it. More often than not, the experience leaves its victims without supportive evidence. Their sense of alienation is profound.

Peter is one of those people who have been forced to live a double life. On the outside he is an ably functioning person with extensive and impressive connections to the normal world. To put it succinctly, he has a life. He is not the stereo-typical alien abductee whose life has been strangled into a fragile form of alien marginalia. Indeed he seems the last sort of person whom one would expect to be coming out with such tales. But Peter has been searching like everyone else who has been affected by such experiences. He was not one to give in to the pressures of marginalization that come with the reception that such claims received from the general public, much of the UFO community, and the more committed skeptics. Peter fought to understand what had happened to him and what was still happening to him. He empathized with others that had apparently had similar experiences, but unlike most other abductees, he set about trying to make a difference.

The 1988 Encounter

Peter Khoury was twenty-four years old. He was living with his Lebanese parents and family in a Sydney suburb. He had been back home for about six months, regaining a sense of comfort with his life. Peter was like the prodigal son returning to the fold. For about six years he had lived a life he is now very happy that he left behind, a rough and dangerous life, moving with the wrong crowd of tough and belligerent people. He grew to see himself as a bit of a "hard head," "a tough guy." He had also been in a difficult relationship. Peter felt that if he had stayed in that life he might well have ended up dead, so he left that life behind and hoped to establish a better one. His family welcomed him back home and he reignited a relationship with his school sweetheart, Vivian. Life was good and heading in the right direction.

Peter was at home that quiet night. At about 11 p.m. he and his father were watching television. His brother Sam had been in Peter's room, sleeping. Sam emerged and asked Peter if he wouldn't mind watching the movie in his room so that he, Sam, could watch it in his favorite chair in the TV room. Peter didn't mind; he went to his room and lay down on the bed. What followed occurred immediately, so it is difficult to link the experience with some sort of sleep-related phenomenon, such as sleep paralysis or hypnagogic imagery.

While . . . lying on my bed, I felt something grab my ankles. As I felt this, a strange numbness, tingling and churning sensation crawled up through my body and right up to my head. I was paralyzed, I could not move any part of my body but for the exception of my eyes, which I could move, open or close. My brain was functioning but I could not do anything physically. I tried to call out to family members but I could not force the words out. At this stage I started to panic thinking I would not walk again. I thought I was truly paralyzed.

As the experience unfolded Peter's first thought was that this was a form of payback, a punishment for the less than satisfactory life he had led for about six years before he returned to the family fold. He thought, if he survived this, that his community would think, God's punishing him, God paid him back. What followed—the encounter with the hooded, three- to four-foot-tall creatures with black, wrinkled faces— changed everything.

The fear was so extreme. I'm gone! I'm dead! It's real! I was petri-fied and paralyzed. Irrational fears were crowding in. They were going to kill me. I could be killed. The fear level was extraordinary. I was stressing out, how could I get out of this?

I became aware of some sort of communication, seemingly tele-pathic, no sound made and yet I could hear the message in my mind. I was told not to worry and I would not be harmed and to relax. As I moved my eyes and looked to the left side . . . I made eye contact with two beings who looked so different to the others. These were thin, tall with big black eyes and a narrow chin. They were goldish-yellow in color.

Astonishingly the stress was gone immediately. The whole fear thing washed away as quickly as a light switching on. How could I be so calm after such a level of fear? My heartbeat was no longer racing. [A whole different situation was unfolding.]

Peter somehow got the impression that one of these tall figures was male, the other female. How he sensed this he doesn't know. Both had an incongruous feature beyond their incredible strangeness. Each had what looked like small, Band-Aid–size "surgical masks." The "female" figure had her little mask on, the "male" one had the mask down. The "masks" seemed to convey to Peter that these strange people were "doctors." Both were wearing whitish gowns, which may have served to reinforce the idea that these beings were "doctors." They also served to highlight the prominent part of their alien anatomy, specifically their faces.

I was so relieved. The one closest to my head communicated with me telepathically, telling me not to worry, [then, paradoxically for Peter, the entity conveyed the thought] *it would be like the last time. The eyes of these tall beings were the source of remarkable feelings. I could feel the emotion through them. It was the eyes that expressed these feelings. You could see the smile in the eyes. Strangely it was like a mother would look at a child. So much love. We are not going to hurt you.*

It was at this stage that I noticed a long needle-like flexible crystal tube. The being then pointed the needle to the top left side of my head and inserted it. A thing on the tip, it's gone in.

It was then that I blacked out.

The next thing I remember I was conscious. I jumped out of bed like a flash, I walked into the TV room where my dad and brother (Sam) were. I noticed they were asleep. I woke my brother up—he looked dazed and lost. As he put it, he felt switched off. I asked him how long it had been since I went to my room. He replied about ten minutes, which was how long I thought it had been. When I left the TV room a film was just starting, yet as I spoke to my brother we realized that the TV station was (apparently) closed and at least one to two hours had passed by.

Neither his brother, who is older than Peter by ten years, nor his father had any idea what had happened. But Peter did have something to say. What he told Sam came in confused fragments and was very strange. Soon it became clear to Sam what Peter was trying to tell him. He told him of floating, apparently, of something touching his head, of people in dark hoods. Sam was initially skeptical, but then realized a good deal of time had passed—from 10:30 or 11 p.m. to something like 2:30 a.m.—and he began to accept that something very strange had occurred. He grew to accept what his younger brother had told him that morning. Sam was very puzzled by the fact that normally he was a very light sleeper. That he, as well as his father, had seemingly been completely "zonked out" for so long was very unusual. He had no way to explain it.

The next day Peter told Vivian, his fiancée, about the strange event.

I explained to her what had happened through the night. As I touched the spot where the needle was inserted, I discovered some dried blood under my fingernail. Vivian took a closer look and noticed a puncture hole and blood. I went to my family doctor and asked for a checkup. The doctor spotted the puncture mark instantly and commented that I must have hit my head on a nail at work. When I tried to explain what had happened, I was laughed at. I had nowhere to go for help, no one to discuss the incident with. It was frustrating to experience something so bizarre, so strange, yet so real.

Rationalizing the Bizarre

Peter was initially anxious and confused about the July 1988 episode. There were physical scars on his body, which he thought were connected to the experience. And he had a possible "biopsy"-style puncture mark on his shin. He was not aware of the world of alien abductions and struggled for understanding. His own Lebanese background and Vivian's Greek family origins provided no comfort. Family members even suggested he had encountered St. Charbel, Lebanon's first officially recognized saint, apparently because of the presence of robes. Such rationalizations provided no explanation for Peter.

Months later, while out driving, Vivian and Peter were shocked to see a gas station billboard poster that featured a tall creature very similar to the ones he had seen. The poster provided no information, just the image. How could this be? What was all this about? Peter thought. The answer came progressively over the next few days. The image they had seen on the billboard was the now familiar Strieber gray alien face. The poster was part of an extended promotional campaign for the paperback edition of Whitley Strieber's *Communion*. Subsequent teaser posters provided more details until the link to the book was made explicit. Peter eventually bought a copy of the book. Vivian read it first, remarking to Peter that he was not going to believe what he would read. To Vivian it was startling. The stories in this bizarre and strange book were oddly familiar. Peter had told her all about them! Upon reading Strieber's book, Peter found it was like a checklist for his own bizarre encounter. Their experiences had several significant details in common—but just as much not in common. However, as far as Peter was concerned, he now had some sort of context to anchor his own strange experience. It was a start. It appeared that his 1988 experience was part of the UFO or alien abduction phenomenon—whatever that was.

During his July 1988 abduction experience Peter Khoury had the strange mental communication *"not to worry, it would be like the last time."* Initially he had no idea what this cryptic comment meant. Nothing in his immediate past bore any resemblance to the bizarre experience that had befallen him. Then he began to reconsider a strange event that had occurred to him in his native Lebanon as a child of seven during the summer of 1971.

He and several other children (a total of six to eight) had gone up onto the flat rooftop of his neighbor's house to play, an everyday activity for them. He recalled that the heavy door leading onto the roof had to be constantly pushed open again as each child went through. Peter was the last to walk through the door. Then he looked up. Gabby, his cousin, was looking up. He then saw all his friends "frozen" like statues in front of him. Above them was a big ball, a helicopter he thought at the time. It was very close. Peter thought he could have been able to reach up and touch this strange helicopter, a silent egg-shaped craft hovering above them. Peter could make out the presence of two tall thin people inside the object, but the light was such that he could not make out much detail. Apparently, all the children were just watching or standing in silence. The children later found themselves on the ground floor after some time had elapsed, with no apparent memory of coming down off the rooftop. Peter was the only child who seemed to have a recollection of seeing the strange object and its occupants. At the time, the only context Peter had for the experience, while logically unsatisfactory, was that it was an Israeli helicopter—a very strange helicopter, at that.

In fact, like many abductees, Peter Khoury has had many strange experiences. Peter's mother told him (in July 2003) about a strange experience that occurred when he was just twenty-two days old. It was mid-May of 1964 and Peter and his mother were at home in the coastal Lebanese port town of Chekka. At about 4 a.m. she was nursing Peter in her arms. She was half asleep when she observed the figure of a man at the window. For Lebanon, this was a rather unusual looking man. Rather than a typical dark-haired Lebanese, what she saw was a blond-haired, very fair-skinned man in a long-sleeved, black turtleneck top, standing at the window. He appeared to have a long handsome face, with his rather long hair parted at the side near his ear. The man seemed to be just looking through the window at them. She had no sense of fear and cannot remember how this strange early-morning encounter ended. Enquiries were made as to whether the stranger at the window might have been a sailor from a ship in port, but ultimately there seemed to be no obvious answer and it remained for Peter's mother a puzzling memory. When he asked her about UFOs over Lebanon, she volunteered that they saw lights all the time and called them "flying plates," but everything like that was put down to the Israelis.

Another puzzling experience that involved Peter's mother occurred at the family home in Sydney, some years before the 1988 episode. She was startled to see an unknown man inside the house walking through a hall doorway into a room. She found the room empty. The men of the house were alerted and a search for the intruder was instigated. There seemed to be no conventional way to account for the presence of the man. The incident became the stuff of family legend—a "ghost" in the house.

The house was the location of Peter Khoury's striking 1988 experience. That experience was preceded by a series of strange events. For a two-week period leading up to the "alien paralysis" incident, the residence was plagued by repeated occurrences of loud footsteps on the driveway. To Peter and his brothers, it seemed like a lot of heavy-booted men were entering the property night after night, usually in the early hours of the morning, and were coming down the driveway and seemingly passing through the closed garage area. The men of the family tried repeatedly to find the source of these heavy footsteps, but their searches always ended in frustration, with no culprits apprehended. The footsteps seemed to defy logical explanation.

These episodes ended with a striking event. Peter heard the footsteps again, but this time he was prevented from getting out to investigate. He seemed to be paralyzed on his bed, which faced the window to the driveway. He became aware that at the window was the silhouette of a man, but for someone to be in that position seemed impossible because the driveway was about eight feet lower than the window. There was no sense of fear, just concern, mainly regarding the feeling of paralysis. Peter couldn't understand what was stopping him. Suddenly the silhouette was gone and the paralysis was gone as well. About a week later the much more severe paralysis of the July 12, 1988, "alien abduction" occurred.

Sleep Paralysis?

Could Peter Khoury's July 1988 alien encounter be an example of the well-established phenomenon known as "sleep paralysis"? John O'Neill, a member of Australian Skeptics, argued that sleep paralysis could explain the paralysis Peter Khoury experienced, hallucination could explain his experience with extraterrestrials, falling asleep could explain his missing time, and "the puncture wound and scab could be from anything—a pimple, an insect bite or anything else in approximately the right location." O'Neill accepted that Peter's experience was genuine, as he himself had had a sleep paralysis episode. But his experience was typically vague, like most sleep paralysis episodes—"the feeling of some other entity being in the room, but [I] did not know and could not see what."[12] That vague description falls far short of the detailed and ordered description Peter Khoury offers of his experience.

Several aspects of Peter's encounter actually defy this simplistic categorization. First of all, he was not asleep, nor had he reached the state that precedes sleep, the hypnagogic state, that often yields fleeting imagery of a hallucinatory nature. Furthermore, his experience has an ordered sequence of events that were recollected consciously; they were not vague sensory experiences of fragmented sleep. Then there are the physical marks, namely the injury to his head, where he felt the strange needle being stuck in, along with the biopsy-like mark on his leg, which certainly can't be accounted for as a classic sleep paralysis episode or as a product of the sleep/awake interface that spawns hypnagogic imagery.

Of course, night terrors, sleep paralysis, and hypnagogic (presleep) and hypnopompic (postsleep) visions do provide a rich range of sleep phenomena that might account for some abduction experiences. Hallucination specialist Professor Ronald Siegel has undertaken extensive research into these areas and describes some of it in his book *Fire in the Brain*. He has had his own sleep paralysis experience, which involved the classic elements of a weight on his chest and a "murky presence," with a "dusty odor," that approached his bed. It communicated with him in a strange way, "almost like English spoken backward." The "presence" then straddled his body, "folding itself along the curve of my back. . . . There was a texture of sexual intoxication and terror in the room." Siegel began to slip into unconsciousness, but then the voice stopped and he had the sense that the "presence" slowly left the room. Siegel concluded that this episode was the product of two striking and often alarming sleep phenomena—sleep paralysis and hypnopompic hallucination—which was mediated by his own "images, thoughts, fantasies, memories, and dreams." A very broad cross section of the population experiences these phenomena. These experiences are not due to psychological problems, but their broad resonances with alien abduction lore require us to consider them as possible explanations.[13]

During 2002 and 2003 research conducted at Harvard University reignited the idea of psychological mechanisms as a possible explanation for alien abduction reports. The key players in the debate were psychologists Susan Clancy and Richard McNally. "False memory creation was examined in people who reported having recovered memories of traumatic events that are unlikely to have occurred: abduction by space aliens," states the abstract to their paper, which was published in the *Journal of Abnormal Psychology*. The researchers examined false recall and false recognition in three groups: "people reporting recovered memories of alien abduction, people who believe they were abducted by aliens but have no memories, and people who deny having been abducted by aliens." They concluded: "Those reporting recovered and repressed memories of alien abduction were more prone than control participants to exhibit false recall and recognition. The groups did not differ in correct recall or recognition. Hypnotic suggestibility, depressive symptoms, and schizotypic features were significant predictors of false recall and false recognition."[14]

This research is interesting but has some severe limitations. The sample sizes used in each group were very small— 11, 9, and 13 respectively. Even more critically, the profile of the "recovered memory" group is flawed. The paper states, "None of the participants interviewed reported continuous memories of alien abduction (i.e., memories of alien abduction that were never forgotten). . . . Memories were recovered both in therapy with the help of certain therapeutic techniques (e.g., hypnosis) and spontaneously, after reading books, watching movies, or seeing television shows depicting such episodes."[15] In fact, there are a significant number of people reporting abductions who do not rely on hypnosis and other therapeutic techniques. Their memories are based on clearly recollected incidents for which they have no memory loss or delayed recall.

The fact that apparently hallucinatory phenomena can occur in states and situations similar to those occurring in alleged abduction experiences begs the question that the details of such experiences need to be considered carefully, rather than superficially rationalized away via token categorization. Could these experiences also be the product of psychological phenomena springing from encounters with an alien intrusion that may even be manipulated by the entities responsible for those encounters? We must exercise caution against either the simplistic acceptance of alien causation or skeptical dismissal. Such uncertainties make it all the more important to concentrate on a broad forensic approach to abduction experiences, but physical evidence, wherever possible, needs to be at the heart of these investigations.

CHAPTER TWO

Hair of the Alien

I AM NO STRANGER TO EXTRAORDINARY CLAIMS IN THE BIZARRE world of alien abduction stories, having quietly investi-gated abduction claims since the seventies.[16] I first heard of Peter Khoury in 1991, and in the years that followed, he re-vealed his story to me, but only gradually, in fragments here and there. I first interviewed him at length on November 1, 1993. He told me about his 1988 abduction experience in Sydney, the rooftop experience during his childhood in Lebanon, and a UFO sighting he had witnessed with his fi-ancée, Vivian, in February of 1988. This latter event took place near the Cooks River, in Hurlstone Park, a suburb of Sydney. For about twenty minutes, Vivian and Peter had watched a strange light jump around erratically in the sky. There would be a flash, or a shot of a light beam, and then the light would suddenly be in a new location. It was like a laser firing. Vivian called it "a war with stars."

Eventually I realized that Peter was being somewhat circumspect with his story. In fact, he did not reveal to me his landmark 1992 experience until four years later. The story slipped out in my presence on July 13, 1996, though its extraordinary significance would not be revealed for some time. At the time, we were meeting with some independent film producers who were interested in my UFO research and Peter's own experiences. Peter was opening up more than usual. Suddenly Peter came out with an abduction story replete with physical evidence of a most unusual kind.

I made a record of it in my journal at the time: "Peter described a highly personal experience he had at home about two to three years ago—sexual—being forced onto a strange female being—onto its breast. He resisted; then a very strong force pushed him against the breast—he bit a nipple—felt something like a piece of rubber caught in his throat for days—he discovered some very fine wispy hair under his foreskin—he still has one small strand, which he keeps in a small plastic bag. He showed this to the two media people and me."

The following account of Peter's 1992 experience is based on a number of interviews I conducted with him in the years that followed. I have edited and reorganized the account to eliminate repetition and to adhere to the original flow of events. It begins with a discussion of some injuries he received on the job on May 13, 1992.

BC: Peter, can you tell me, in your own words, what happened?

PK: *Well, basically it was at a time I had had head injuries and I was on a lot of medication, and I was pretty sick.*

BC: What were the head injuries about?

No Hallucination

PK: I got attacked at a job site by three guys. I was hit with shovels on the head; trowels were thrown at me. So I had pretty bad injuries. I suffered a lot of pain. I was on a lot of medication. Because of that I would vomit constantly, particularly in the morning. I know there was one time I vomited about ten times, and that was just while driving Vivian from home to the station, which is about three minutes away by car. I had to pull up about ten times. I would just get sick.

If I didn't have what I have as proof for me, I would say that I was on enough drugs, pain killers, et cetera, to hallucinate maybe.

BC: What sort of things were you on?

PK: *Panadeine Forte* [paracetamol and codeine phosphate, a pain reliever].

BC: Not generally known for their hallucinogenic properties.

PK: *I was put on Prozac, but I only took one tablet. I read up on it and I didn't like what I read.*

BC: Were you suffering from fevers?

PK: *No just a lot of headaches, pain, severe migraines, a lot of vomiting, a lot of dizziness. But no, I wasn't hallucinating. . . . I had been on medication all along up until the experience and even after. I felt around that time . . . I was being given too many different medications, so I approached the doctors and narrowed it down and just basically started taking Panadeine Forte and Voltarin . . . for six to seven months after that. When I had extreme headaches I would take up to three Panadeine Forte and that's really it.*

We can be reasonably certain about the date the strange intrusion into Peter Khoury's Sydney suburban home. He recorded a brief note in his diary as follows:

> *Thursday July 23, 1992:*
> *Had a very weird experience this morning with two fe-*

males. 7.00 a.m. after dropping Viv off at station got really sick a few times went straight to bed. All of a sudden two females appeared from nowhere. A blonde and a Chinese girl. Very weird looking eyes and color of blond. Both were naked. Blonde push me to her breast a few times then I bit her nipple and started coughing. She showed no emotion, blood or screamed in pain. I went to toilet and found two hairs under my foreskin. I put them in plastic bag. It is in my filing cabinet.

These few words do not convey the striking nature of Peter Khoury's strange experience, as my interviews with him revealed.

PK: *I'd had a shower the night before. Got up in the morning, drove my wife to the station. I did pull up a few times and get sick. I came back home. I felt really ill, so I went straight back to bed at about 7:05 a.m. I was clothed.*

BC: Can you just describe what sort of clothing you were in at the time?

PK: *Tracksuit pants and a sloppy joe . . . and just jocks . . . my underwear was on. . . .*

It took a couple of minutes. I fell asleep. At about 7:30 I sat up bolt upright in bed. I don't know why. There was no reason for it.

It was a feeling of something, more so like if something light stepped onto the bed.

BC: Well, let's get back to your sense of awareness of lying on the bed, and then suddenly upright. I think you mentioned to me before there was this sense of either something going onto the bed, like a cat jumping on the bed.

PK: *Yes. Maybe, that's what woke me up. It was a feeling of something. Yes, you're right. There was something, more so like if something light stepped . . . [it is] a normal, conventional bed, base ensemble, with a mattress and bed head . . . a very firm bed. If some-*

one sat on the bed, you would notice they sat on the bed . . . and maybe that's what made me sit bolt upright—the movement, feeling something on the bed.

My eyes opened. I noticed that there were two females on the bed, but for two women to be on the bed it was too light, the movement was too light.

What is going on? What's happening here? I knew I was looking at something that didn't belong in my room or in my house, but there they were. How the hell did they get there?

BC: Had the front door been locked?

PK: *Oh yes, the front door would have been locked.*

A Strange Viewpoint

PK: *At the time I was still sort of waking up. In that state of trying to wake up, trying to put my senses together, I was bewildered a bit, more shocked and confused, but there was no fear. I was trying to get my bearings.*

I was looking at a transparent image. I'm looking at myself. Then I caught up! This only lasted a few seconds.

The way I was looking at it, was like I was looking through the back of my head, like through my eyes, sitting back behind me, watching myself, like I could see myself in front. Like I could see the back view of me. . . . It is really hard to explain, but I was virtually looking, [as if] standing back, looking through my image in front of me. I was watching this and I could see myself as well as the two women. Then I caught up—my physical body.

I was now myself looking at the strange scene before me. It seemed very real. [Ultimately it would be all too real!]

Although I had been involved in the UFO field and I had come across a lot of cases, I don't think I had come across a case where an abduction experience had occurred during the day or morning hours—

seven o'clock in the morning. It has always been nighttime or early hours such as three, never during the daylight.

Strange Intruders

PK: *As far as the description of these women, one of them was blond and the other was dark-haired and oriental-looking.*

It would be very hard for me to find two women who looked like these ones, especially the blond one.

Well, I looked at the blond one. I was trying to analyze how they got there first. Like, someone's broken into the house. It was just such a shock to me to wake up to see that.

BC: How are these two woman seated on the bed?

PK: *Well, the one that was opposite me, directly opposite me. It was on a double bed, so there was a lot of room. The one opposite me was a blond-looking woman and she was sitting with her legs tucked under her backside. She was virtually sitting on her calves. You could imagine her kneeling down with her legs folded, sitting down. The one on the other side was kneeling halfway. She wasn't actually sitting down on her calves. She was sitting upright a bit.*

The Blonde

PK: *I'll describe the blond one first. She was . . . in her mid-thirties; she had good skin; the hair was pretty weird, for the morning anyway. You don't see a woman get up with makeup and her hair all done up. This one, her hair was done up like the wind was hitting it, like it was blown back. It was different. I had never seen a hairstyle like that; it was curled something like Farrah Fawcett, but to an extreme. . . . It just looked really exotic in a way. . . . It was blond. Her facial features . . . she had protruding cheeks, very high cheeks. The nose was long; I wouldn't say too long, such as a big nose; but long as in* stretched, with the face, because it was a long face; a longer face than I was accustomed to in females I know. Her eyes were two to three times bigger than our eyes. I looked at the eyes and I knew I wasn't looking at a human female. . . . Not forgetting that I had previous experiences . . . I connected it straight away, to something of that kind. She looked humanoid, and human in features. Her mouth, her lips were normal in size.

BC: What about the shape of the face?

PK: *The shape of the face was longer than usual. It was as if it was somehow stretched . . . just longer, longer than ours, pointier than ours. . . . The head just didn't look right. It was longer and narrower than a human being, for example. Her body . . . had normal female breasts. She was naked. . . . I didn't notice anything different about her body. . . . She had average-size breasts, well proportioned. The only thing that looked different to me was the face, in that it was longer than ours, the eyes were two to three times bigger than ours. . . . The chin was pointier . . . like a witch's long chin in movies. . . . The hair covered the ears. . . . Her hair came down to about halfway down her back, and like really high up, it was sitting up. The hair looked really exotic. . . . It looked wispy in a way; although it looked nicely done, it looked to me like it was frail in a way . . . it was flimsy . . . it wasn't very strong hair.*

The "Asian" Woman

PK: *The dark-haired woman looked Asian. . . . The Asian one, she would be average height, five feet eight inches or so. Her features also weren't completely human—the cheekbones reminded me of Asian women, but too extreme, it was as if she had been punched in the cheeks by Mike Tyson or something . . . the cheeks were too puffy; the eyes were too big once again, about the same size as the other one['s]; her eyes were dark, almost black; I don't remember seeing white in the eyes. . . .*

You get Asians, and they would say the same about us, but they all seem to look alike. If I had a lineup of ten Asian women, and she was in the lineup, I would say, yes, I would pick her out, because she had distinctive features about her. . . . The eyes and the cheeks . . .

The blond one was directly opposite me, and when I sat upright she was probably two feet away from me. The other one was to the right and sitting on the side of the bed . . . facing us from the side. When I looked at her I got the impression that she was watching the blond one and learning how to interact. She was just there with this concentrated-like look. . . . She was looking at both of us . . . at what she was doing and at the reactions, my reactions basically to the blond woman.

BC: What about the musculature? Were they thin boned or were they muscular?

PK: *Well, I got the impression from the blond one that she was very tall because from the waist up she was taller than me. When she sat up . . . when she cradled me and brought me to her breast, I would say she had . . . a head and a half higher than me. . . .* [Peter is 182 centimeters, or approximately six feet, tall] *I would say she was a lot taller than me, that's for sure. The blonde had light-colored eyes, maybe bluish. . . . She had normal human-looking eyes except for the size. But the other one, it was like looking at a TV screen, that tone of dark.*

BC: What about skin tone?

PK: *One was light* [the blonde]. *She was like a normal westerner; the tone of the skin was very light. The other one had darker skin, darker than an Asian woman, maybe like . . . from India, that darker tone of color. She had straight black hair down to her shoulders, super straight. . . . I didn't see her hair move; when she would look at us . . . it was sitting there stiff. It looked like a veil, like a headset. . . . It looked like it didn't fit there. . . . Her skin tone was very dark. Looking at the blond one and looking at her, and looking at these little differences, you knew you weren't looking at a human female. . . . The Asian woman's face looked more human than the other one, except for the eyes and cheekbones . . . they just sat up too high. They just looked like a deformed human. . . . The blonde very definitely . . . I have never seen a human looking like that. . . .*

He did not notice any underarm hair, nor did he have the opportunity to see if there was pubic hair. The blond woman's nipples were quite prominent and the "Asian" woman's were quite small.

Extraordinary Reactions

PK: *[The blonde] just basically reached out and grabbed me from the back of the head with both her hands. She cupped the back of my head and forced me towards her breast, towards her left breast. I resisted and she forced me again. I resisted and a third time that she forced me, pulled me towards her. She was pretty strong . . . when I'd resist, she would pull me straight back with ease. . . .*

She was strong, but she was so gentle the first time, almost motioning me towards her. I pushed her away. Then she used a bit more force. I used a bit more force to push her away, then the third time she used a lot more force. At first I couldn't push her away.

She had a strength that belied her appearance. In fact, Peter felt she was stronger than him. Peter is well built and, working in the building trade, is in no sense a weakling.

PK: *You know, I didn't feel any weight on me. It seemed to be controlled strength. She was strong. My reaction was, don't fight, be calm. It was a surprise that someone so frail-looking—a woman—was so strong.*

She pulled me over and my mouth was . . . on her nipple and I bit.

BC: Why did you do that?

PK: *I don't know. I've asked myself many times. I know how much it would hurt a woman if you did bite her nipple. I did that, and I took a chunk, a little bit off, because I swallowed it and it was stuck in my throat for three days.*

I don't know why I did it. I don't know whether it was like a defense thing for me, to get away from her. I know there might be a lot of people who will say this is a sexual fantasy or whatever. I've been around. I had been with two women. It wasn't a cosmic experience for me. It was normal. I haven't done it again. I didn't think it was that great. It's just normal, nothing to write home about. . . . I think that my biting her nipple was the only way I could say that I didn't want to do this.

The thing is that when I bit her nipple . . . she didn't scream, she didn't cry, she didn't react at all in pain. No way at all was there any pain associated with what I did. That really, like, put a big question mark on what was going on with me. Like, what the hell was going on? There was no blood, there was nothing, no trace whatsoever. It was as if I took a bit out of a plastic dummy or mannequin that was made of rubber or something.

I suppose when I bit, that helped me push her away . . . the expression on her face was like, this isn't the way. In a way, it was shock or confusion. Like, this isn't the way it's supposed to be. She looked at the Asian one, I remember, her looking over. They looked straight at each other's face and looked at me like this isn't the way it's supposed to happen. You've done this wrong.

The blond one was showing the dark-haired one how it's supposed to be done or how to interact or . . . whether it was a sexual interaction . . . I don't know what sort of interaction we had. . . .

BC: Getting back to the two women on the bed, you said the dark-haired one was looking as if she was being shown what to do, what was the right thing to do. What happened then?

PK: *The expression on her face was blank. It was if she was there . . . you know when you look at someone and you see a sparkle in their eye, you know they're alive . . . I didn't see that in both their eyes. It was just like looking at someone with glass eyes. . . . There was some-*

thing missing. . . . The Asian one in particular, her stare was just straight at us, just watching, analyzing what's going on. The other one, her actions, as far as I was concerned, were clinical . . . no emotion whatsoever.

They knew what they were doing. They were there for a reason. . . . The first thing that came into my mind was babies. That may have been due to the Asian-looking woman having a little bit of a pot belly. They were there for a reason, to fall pregnant, I thought. . . . I don't know . . . [Peter also suggested that her more prominent belly could have been accentuated by the way she was sitting.] *It was so controlled. There was a reason behind it, [someone or something] made it happen.*

A Vanishing Act

PK: *I'm swallowing and there is this thing stuck in my throat and I started coughing, and as soon as I started having this coughing fit, that's when everything stopped, they weren't there anymore. I got out of bed.*

BC: How do you mean?

PK: *They just vanished. One second they were there. When I bit the nipple, it felt to me as if I had bitten a little bit of elastic band, for example, rubbery substance. As soon as I did that I saw the expressions and I started to cough, got this coughing fit, and I might have taken my eyes off them for a split second, I think, then I've looked up and they weren't there anymore.*

BC: Were you aware, between the time of coughing and being aware that they weren't there, if there was any sense of, like, continuity of time?

PK: *Oh yes, I don't believe I blacked out. I don't think so.*

BC: What about this sense you described earlier of having viewed most of this through the sense of being behind yourself and seeing through you?

PK: *Well, that's what I mean. Soon as I had this coughing fit, I've looked up and they were gone and I was looking through my own eyes. There was nothing there. It was just me on the bed. But the whole time when this was happening I had the feeling as if I was . . . looking through the back of my own head and through my own eyes . . . as a second person looking through. It was really weird. It was like looking through binoculars, but through the back of my own head. I was watching myself . . . having this experience. . . . I don't know how to understand it myself, it's pretty weird. . . . It gets . . . you know, being involved in the field you would think you would have some answers. It's ridiculous.*

I got the impression that this sense of transparent viewing lasted the whole experience until Peter started coughing. But Peter said this was not correct, that the duration of the transparency effect was very short-lived, only lasting from when he awoke until he rose from the bed. When he was upright, facing the blond woman straddled across his upper legs, his viewpoint returned to normal.

PK: *I got out of bed coughing. All this time I had really bad coughing. I had something stuck in the back of my throat that I was trying to cough out. I remember getting a glass of water, having a drink. Didn't do anything. [I] walked straight to the bathroom to use the toilet.*

BC: Why did you go and drink some water?

PK: *Well, to wash this thing down. . . . There was definitely something stuck in my throat.*

While I was home on my own, I was coughing. I was trying to drink. But after the first glass of water I went into the bathroom to use the toilet.

BC: Why was that?

PK: *I had to go to the toilet.*

BC: Would that have been normal at that time?

PK: *Not really. . . . It didn't seem odd to me. . . . I had a strong urge* to go to the toilet. I had a glass of water and that even made it more so that I needed to go. I went to the toilet. As I said before, I had a shower the night before. Then I went straight to bed. Vivian and I just fell asleep. Nothing happened.

Bizarre Discoveries

PK: *Basically I was in that much pain when I tried to use the toilet. I thought my penis was virtually like slashed. It felt like it was cut up and it was just burning. It was just too painful. I pulled the foreskin back and wrapped around the head and part of my penis was this hair that was wrapped right around it, really tightly wrapped, and there was another little hair that was also intertwined. . . . It really, really hurt. . . . It was so painful. It was like a nylon string, and it was going to cut me. . . . I untangled it. It was really painful to untangle. . . . Not that it cut me, though. But it felt really, really painful to take it off. When I eventually took it off, I came into my office and grabbed a plastic sachet bag, put it in it, [and] sealed it. The reason I did that was because I knew that there was no way, no way at all, that a hair that size and wrapped around the way it was, should have been there, and thinking of these women, the thing in my throat, the hair, something bizarre had just happened and I'm glad now I actually kept the sample, because as you know I've kept it for quite a few years.*

The hair was about ten to twelve centimeters, the other about six to eight centimeters. . . . It was very frail looking, whitish in color, rather than black or blond. . . . It reminded me of a very thin fishing line . . . it wasn't thick. I took a sample of my wife's hair to look at and a sample of my hair. I had the sample in the bag, and . . . it was a major difference. [When I first unraveled the hair] I noticed it was blondish, whitish hair, it didn't even look like my wife's hair. . . . I connected it straightaway to the blond woman. There is no doubt that that hair came from her. How it got there, I don't know. I've got no idea. . . . Vi-

vian's hair was thicker and way darker in color. We even tried to find the lightest hair, even white hair on her. . . . It wasn't anything to match hers. . . .

How did it get there? . . . Because I had no memory of anything happening. But having a woman on your bed who is naked . . . who pulls you over to her breast, trying to instigate something . . . maybe I was a fool to bite the breast. . . . You handle situations on the spur of the moment. Maybe I should have done it differently. But that was my way of dealing with it, to put a stop to it there and then. . . . It was so clinical. . . . They had no feelings, no emotion. . . .

With regard to the placement of the hair on his uncircumcised penis, Peter said:

The only way it could have got there was if somebody had actually pulled the foreskin back, wrapped it on there and left it that way . . . and I can't see anybody doing that when I'm asleep. . . . I wished I had a recollection of how it got there. . . . It wasn't knotted, just twisted . . . it was wrapped around it in a coil. That's why it was so painful. It wasn't on just one area, wrapped around. It was hurting in a few places, not just one spot. . . . [The hair, when taken off] didn't curl up. It was so flimsy. It had a spring to it, but not curled. It wasn't straight . . . like a piece of string. . . .

Peter felt no pain in his penis until he went to the toilet.

He estimated the experience lasted about five minutes, perhaps a little longer. There was no verbal communication throughout the experience. There seemed to be no telepathic communication either, but he knew that the women were communicating between themselves, particularly conveying a sense that the interaction that had gone on between Peter and the blonde "was wrong, don't learn this."

There was no sound during the incident. The bedroom curtain was drawn. Though the room was dark, there was sufficient illumination to clearly make out things in the room.

Peter did not remember if there was any reflection from the wardrobe mirror. He did not notice. His attention was focused on the two women, particularly the blonde, who was very close to him.

Peter does not remember feeling sexually excited during this episode. There have been innuendos that there must have been a more explicit sexual event to account for the hair being wrapped around his penis. He can understand where people might get this idea, or that he might have had the biggest erection, or the best sex of his life, but he stresses there was none of that. He doesn't remember an overtly sexual aspect to the bizarre experience. In hindsight, he feels he handled the situation inappropriately, but he in part attributes his reaction to the shock of the episode. Peter was unaware of any similar encounter in the wider UFO experience, although he stressed that he was relatively unaware of the extent of the phenomenon or the literature.

Fractured Continuity?

BC: Given the set of circumstances that you actually recollect, how do you connect this with the women, because you have described the sense that there was no sort of break in continuity, they were there, then you looked up, and they were gone. Then there was this sense of this hair there.

PK: *I don't know if something happened before, and when I drove Vivian to the station it [then] continued [when I got back]. I don't know if it happened as soon as I went to sleep.*

BC: When you had this awareness, this encounter with two ladies, after it ended and you were coughing, were you aware of your clothing being messed about or was it just as if you were dressed the same way as you went to bed?

PK: *No, I was dressed the same way I went to bed. The only thing is the coughing didn't stop. . . .*

BC: Had you gone to bed under the sheets?

PK: *Yes.*

BC: Were you aware whether you were under the sheets when you . . .

PK: *When I sat bolt upright, actually, no, I wasn't. I was on the bed, on top of the bed. . . . My legs were exposed. I could see myself.*

BC: Where were the sheets at that time?

PK: *Under me. . . . That's interesting actually. I never thought of that.*

While this might suggest some sort of discontinuity in the progression of events—it began with the sheets over his lower body, and ends with them under him, with no awareness of how this took place—Peter has no memory of anything else happening.

Immediate Aftermath

BC: What did you do about it at the time?

PK: *Virtually nothing. I was more concerned about this thing stuck in my throat for three days. . . . It was so annoying. . . . Somehow I wished I could have spat it out. . . . On the third day I stopped coughing and it was gone. I knew that something happened that wasn't ordinary.*

I had a coughing fit that went for hours, for three days. I tried bread. My mum called a couple of times during the day and heard me coughing over the phone. Just eat some bread and wash it down. Not that I told her what it was. I spoke to Vivian on the phone.

BC: When did you speak to her?

PK: *It would have been half an hour after it happened. . . .*

BC: What did you say to her?

PK: *She would have just gotten in to work. Probably about 8:30 I spoke to her. . . . She heard me coughing, and I said I've got this thing stuck in my throat. I've been trying to wash it down. It won't go. It's annoying. I said to her when you come home I just want to talk to you about something and left it at that.*

I waited until she came home.

I recall when Vivian came home, I said to her something happened, I'm not ready to talk about it yet, but when I mention to you about the coughing fits I'm having now, if I tell you in a week, a month, whenever I tell you, connect the two. When I say to you remember when I had that coughing fit . . . relate [it to] what I'm talking about. Yes, she said, no problem. And, it took me two weeks.

Peter's own diary entry indicates that he actually delayed three weeks, namely: *August 13: "females experience—talked to Vivian about the coughing fit I had for three days. I told her what happened."* He showed her the hair.

BC: Why?

PK: *I don't know. In a way I felt guilty, then I felt I have no control over what happened. And then at the same time I thought, how was Vivian going to deal with this? I mean . . . we were married, we didn't have kids at the time. . . . Like, how am I going to explain it to her? How is she going to feel? How is she going to deal with it? Is she going to feel threatened? She surprised me actually. [Three] weeks down the track I said to her, look, remember when I had the coughing fit. She said yes. I said, like, this is what happened and there were these females on the bed, the hair, etc., the coughing. I said, what do you think about it? She said, it's not like something you can control, it's not like you're inviting a woman over while I'm not here. What can we do about it? I was shocked. She accepted it better than I did.*

CHAPTER THREE

Seeking Help

PETER KHOURY FOUND THAT GETTING ASSISTANCE TO UNDER-stand his bizarre experiences, beginning with the 1988 episode, was difficult at best. Professional help was not forthcoming. He even contemplated hypnosis, but the costs were daunting; besides, as soon as he mentioned what it was about, the hypnotherapists didn't want to see him. Then in November 1991, he heard about a local UFO group called UFO Research New South Wales, which was about to have its inaugural meeting. Peter was seeking to understand what had happened to him in 1988 and joined the group. The puzzling encounter with the two females that would yield the hair sample was still some months away. Peter had entered the surreal world of ufology.

The UFO group is a beacon for people in all walks of life who are attracted to the mystery of UFOs and aliens. Some groups are well organized, highly focused, and serious, while others are downright bizarre and irrational. Often, UFO groups are a mixture of these two competing factions. Many try to attract public membership and attention, while others prefer a low-profile approach that favors networking, research, and investigation. Wild speculation and political infighting is often rampant.

UFOR (NSW) was not immune to these issues and problems. Driven by the need to understand his own experiences, Peter found himself the group's de facto "abduction coordinator" and tentatively developing their abductee support subgroup. But what he found was a group that was not really ready to come to grips with the alien abduction subject. Many in the fledgling group saw alien abduction reports as a can of worms. One of the coordinators of the group described abductions as "just too sexy, too sensational, and 'not nice.'" One can perhaps understand this sort of conservative response from a newly formed civilian group trying to focus on UFOs and struggling to achieve and maintain respectability in the eyes of the general public. Alien abductions were simply too controversial. Peter's support group was a "bastard child."

This was the formative period of Peter's missionary zeal for the cause of accepting peoples' claims of alien abduction experiences. Although Peter's traumatic 1988 experience had clear resonances with the experiences of people like Whitley Strieber, whose written accounts had been read by millions of people around the world, Peter began to realize that there was, at best, a limited tolerance and a narrow understanding of his experience, and that in general people regarded the issue with skepticism, and sometimes with hostility and rejection.

An Implant?

Peter Khoury clearly remembered that a needle-like device had inserted something into his head during his 1988 encounter and had left behind a "puncture mark." He believed that perhaps now he had an implant in his head. Nearly four years later there was still a noticeable lump at the site of the puncture mark. The doughnut-shaped lump seemed to grow intermittently to about half the size of a marble, and then shrink, but it would never entirely disappear. He occasionally worried that it might be a tumor, but as it diminished in size, so did his concerns.[17]

Three days before his CAT scan for the head injury he received on the job in May 1992, Peter was scratching the site of his puncture mark as he sat with Vivian in the evening. He had come to note that there seemed to be a very thin "wire" sticking out from the spot. Peter and Vivian could feel it, but they couldn't see it. Now it had begun to irritate him. He continued to scratch the area. Then suddenly he felt something pop out of the spot and he caught a glimpse of something fall between him and Vivian. She didn't see it, but both heard it hit the coffee table and apparently bounce into the fibers of the shag carpet. Peter thought it looked dark brown and was about the size of a match head. Realizing that he may have dislodged something that had been there since 1988, Peter quickly got a small portable vacuum cleaner and, with a tissue covering the inlet, began to vacuum the area. But he found nothing, and eventually they gave up looking for it. In the wake of that episode the transient lump on his head was now gone, and the CAT scan that followed for the job site assault revealed nothing in the area of the puncture mark.[18]

Confronting Abductions

Dealing with his abduction experience of 1988 was hard enough on Peter Khoury; the situation merely got worse and more difficult after his July 1992 experience. His diary entries reflect his frustrations and concerns. An entry on August 26, almost two weeks after he had told Vivian about it, notes a *"lack of help from committee."* The very next day Peter records, *"do not discuss experience,"* referring to the July 1992 event, because of the lack of positive feedback or interest.

It took the visit of American alien abduction researcher Budd Hopkins in October 1992 to shake the conservatism of the UFO group. Hopkins's public lecture and its impact helped legitimize or popularize the reality of abductions and the UFO group's negative attitude toward the subject began to change. The group now wanted to take over the abduction support group that had been nurtured principally by Peter Khoury.

At a public meeting of the UFO group on December 6, 1992, which I attended, Peter Khoury and his primary associate Jamie Leonarder formally introduced the support group concept. Leonarder had joined the group early in 1992. He was a welfare worker with a Sydney welfare agency that provided support for down and out street people. These included those with drug and other dependency problems, destitute individuals, and other people on the fringe of society. Not surprisingly, Leonarder was attracted to the issues and concerns of people like Peter Khoury—the legion of the abducted. He had a natural empathy for these people, who often had an intense sense of alienation both with mainstream society and the alien realm they seemed to have intersected with.

The support group had its first meeting on January 31,

1993. Peter mentioned the blond woman and the hair sample at a subsequent support group meeting, on February 28, 1993, but because of the earlier lack of positive feedback, he did not show the hair sample at the meeting. Instead, he focused on his 1988 experience, which conformed to the then accepted general form of alien abductions.

But problems quickly escalated for the fledgling support group. Leonarder was concerned about the involvement of nonprofessionals in the group and the dubious dynamics that emerged. Peter stayed with the group for about a year, until the pressures and petty politics began to take their toll. He had received no help in understanding his own experience and felt that the group members were not sensitive enough to the complex needs of those who had gone through such episodes.

A New Support Group

Peter Khoury resigned from UFOR (NSW) at the end of March 1993. Many in the support group left with him, including Leonarder, and together they formed their own group, the UFO Experience Support Association (UFOESA), which was formalized on April 14, 1993. UFOESA, based in Sydney, went on to become one of the most notable manifestations of the support group concept in Australia. It was an impressive beginning, emerging out of frustrations with the formal UFO group structure and describing itself as "a nonprofit, voluntary organization dedicated to helping witnesses and experiencers of UFO events cope with and understand their encounters."

Khoury remarks, "I ask you all to think about the situation abductees are in. What if you became a victim of the same circumstances? Wouldn't you want someone to listen to you and feel confident in the fact that the people you are reporting your experience to understand and support you? I ask you once again to be open-minded to the fact that there are many individuals experiencing this phenomenon throughout the world."

It takes a lot of courage to "come out" and describe events like Peter Khoury's 1988 and 1992 experiences. It takes strength of character to go a step further in trying to help others in similar circumstances. Peter hoped that the team of professionals—including scientists, doctors, psychologists, and therapists—that became part of his UFOESA group would be able to respond to the abduction problem in a way the former group never could. Despite the best of intentions, however, the professionals, with their caution mediated by science and commonsense, questioned the more bizarre claims and used them as criteria for accepting or rejecting the validity of an experience.

For Peter Khoury this issue had suddenly and overwhelmingly become a potent and critical problem. His own bizarre experience of July 1992 with the two strange females told Peter that the strangeness of a claim should not be a reason for automatically rejecting someone's credibility. Early in 1993, he used his own experience to inform the professionals within the group of this lesson, but he was less than convinced that his group of professionals had embraced his suggestion.

Peter told them the basic details of his strange "home invasion" by two "alien" women, the bizarre interactions that followed, and the hair sample he had recovered. Though they listened, to Peter it seemed mere tokenism. From that moment on he learned to be more circumspect about sharing the more bizarre episodes of his unfolding alien abduction experience. If UFOESA's own professionals could not come to terms with the stranger dimensions of the abduction experience, Peter realized there would be little hope for help from outsiders. Peter's 1992 experience became a suppressed but important "sleeper" case biding its time.

An Outsider Enters

Given Peter's experiences with his own group, it hardly seems surprising that he initially chose not to share details of his startling 1992 experience with me. I should note that since getting to know Peter Khoury since 1993, I have found him to be a friendly and reliable person. He has at times become passionate and volatile about his cause and does not tolerate those he feels dishonor the integrity of people who genuinely feel they have experienced an alien abduction. He remains open to whatever these experiences may turn out to be, but insists on the serious and ethical research and support of abductees.

Because of the significance of the hair evidence, I needed to establish with some degree of certainty that the hair was indeed procured in the way Peter described during his July 1992 experience. After he revealed the story of the women to me in 1996, he said that he had made mention of the experience in his diaries, but because of renovations and building at his house at the time, he was unable to find them for several years. He finally found them in April 1999. The diaries confirmed that the incident had indeed occurred in July 1992. Both Peter and Vivian were certain that the incident had occurred after his severe job site head injury of May 13, 1992. Both recalled that the incident with the women and the hair had occurred at a time when Peter was still affected by the injury, particularly with nausea, headaches, and vomiting. It was a very difficult time for them, as Peter could not work.

The diary confirmed that the hair incident actually occurred on July 23, 1992. It also shows that he finally reveals the experience to Vivian three weeks later, on August 13. Peter pointed out that while he was reasonably certain of the diary date, 1992 was a turbulent year for him and his family, and therefore he couldn't be absolutely certain that he made his July 23 entry on the right date. He made this qualification about the diary entry, as in his normal building job diary entries he had occasionally found himself making entries on the wrong page. If he picked these errors up, he would cross them out, make a note, and enter the material on the correct date. So, while not absolutely certain, Peter is pretty confident about the date of the episode.

Other than Vivian, Jamie Leonarder is the person most familiar with Peter's 1992 experience. Jamie's association with Peter was established by the middle of 1992, and given the empathy they had, it was only natural that Peter would tell Jamie about the experience and showed him the hair. Leonarder recollects that this was soon after Peter had revealed his experience to Vivian. Again this anchors the experience around the middle of 1992 and is thus consistent with the July 23 date.

Though a rift would eventually develop between them, I was surprised how much Jamie supported the validity of Peter's July 1992 episode. He told me that he certainly recollected Peter's telling him about the episode and being shown the hair. Jamie was certain that the details of the women and the hair were there from the very beginning and he felt the account was consistent with the story Peter has told many times since. Given the bizarre and implied sexual nature of the incident, Jamie had suggested that perhaps it would be wiser for Peter to be careful about who he told this story to, but to ensure that the sample was kept in a safe place, as it probably was important. Peter's subsequent experiences with his own support group committee would confirm to him the soundness of this advice. I talked to a number of these group professionals, and most had some recollection of Peter's telling them about the experience.

Given these various points of confirmation, both direct and indirect, I could be pretty certain about the July 1992 date of Peter's experience, and that his story about the strange women and the recovery of the strange hair sample had indeed occurred at that time. This very strange episode seemed to have a solid foundation.

CHAPTER FOUR

Alien Impact

WHILE THE EXTRAORDINARY POTENTIAL OF PETER Khoury's 1992 experience would not emerge until the DNA analysis took place on the found hair sample, another very striking case came to my notice the same year I began researching Peter's alien experiences. These two cases would provide me with the kind of reality anchor points I was seeking as a scientist—tangible physical evidence that alien abductions might actually be happening.

In my own research into Australian abduction events, from my earliest work in the seventies until the early nineties, I basically came to hold the position that most abduction cases probably told us more about the human condition than they did about UFOs themselves. I found the majority of abduction cases conspicuously devoid of compelling physical evidence. My science background always made me aware of the fundamental position that extraordinary claims require extraordinary evidence. However, such evidence also often requires extraordinary investigation and research to obtain it.

Compelling evidence for the physical reality of UFOs comes from physical traces in UFO landing events, so-called EM (electromagnetic) effects in close encounter incidents, particularly car stalling type cases, radar sightings, physical effects on witnesses, and to a lesser extent photographic evidence.[19] But similar evidence to support the reality of abduction events has been lacking or is relatively unconvincing. And yet such events have come to dominate the entire UFO scene. Indeed, it seems as if the UFO phenomenon itself has been abducted by the alien abduction phenomenon.

One of the few compelling exceptions to the ambiguity and uncertainty that pervade the alien abduction experience is the 1993 Kelly Cahill case. It proved to be a groundbreaking case because it involved an abduction that was observed by three groups of apparently independent witnesses and was supported by an intriguing array of physical evidence. This event represented a turning point for me. The physical evidence makes it difficult to dismiss this case. Actually it became a siren call for me to reevaluate my preference for exotic and misunderstood prosaic vagaries of the human condition as explanations for alien abduction experiences. Ultimately, this episode was not without its frustrations,[20] but along with Peter Khoury's 1992 experience, it would help arrest my drift away from accepting the possible reality of alien abduction stories. Each of these two cases were powerful touchstones for the argument that maybe alien abduction stories should not be dismissed as aberrations of an all too human kind.

Kelly Cahill's Encounter

Kelly Cahill first contacted me back on October 4, 1993, seeking assistance in understanding a bizarre experience she had near the outer Melbourne suburban housing development of Narre Warren North, in the foothills of the Dandenongs, Victoria, between Belgrave and Fountain Gate, during the early hours of August 8, 1993. To conduct the follow-up investigation, I referred Kelly Cahill to John Auchettl and his group Phenomena Research Australia (PRA).

Kelly's account is extraordinarily potent and compelling. I will quote here from an interview Kelly gave to my associate, Robb Tilley, on March 21, 1994, as the details of the incident that emerged during that interview are consistent with those she initially and later related to me, with what she told John Auchettl of PRA, and with her own written account of the experience.

The Incident Prelude

My husband and I were driving to my girlfriend's place up in the mountains. It was her daughter's . . . birthday.

It was just after dark and we were nearly there, about half an hour from her place. It takes us about an hour and a half from our place. It was just after seven.

[The area has] little bits of field . . . and then you run right into a major shopping center . . .

My husband drives really fast. . . . I'm just busy looking out the window. It's turned dark and I look over towards this field as we are going past and I see a ring of orange lights. It was the first time I ever thought I had seen something that wasn't normal, you know what I mean. I was going to shut my mouth. I thought, "No, he's just going to have a go at me." But a couple of minutes up the road I said, "I swear I saw a UFO." He said, "Don't be stupid! It was probably a helicopter." I said, "It wasn't making any noise. It was just sitting on the ground." Anyway, after a few jibes at me, he forgot all about it, and we arrived at [my girlfriend's place].

When we were there, my friends bring up this conversation, about what I thought I had seen. Her father says, "You think you've seen little green men or something, Kelly?" and all this sort of stuff. It was turned into a joke, and I just totally forgot about it.

We went out and played Bingo. We came back. We had a bit of a problem about what time we left. As far as my husband was concerned, because we got home at 2:30 in the morning, it means we must have left at one o'clock. But that night, I think we left at a quarter to twelve. We got back from Bingo at about eleven, and we didn't stay for very long because [my friend's] daughter's boyfriend had just . . . split up with her and had gone home with the new girlfriend, you know what I mean . . . and she was really upset and crying, and we didn't want to stay. So we weren't there for that long at all.

The Encounter Begins

Anyway, we were driving back down the road in the same stretch. Both of us, just me and my husband . . . we both saw this, mind you . . . in front of us, hovering above the road. It was just something sitting there. . . . I couldn't tell what it was. We were at first far away, but as you got closer to it it was sort of . . . well, it wasn't like the orange light in the field, it was a round shape with some sort of glass around, or what looked like windows and lights around the bottom. Because it was dark, you couldn't really tell at first. But as we got closer and closer, there was no noise or anything.

Even my husband's going, "You're right! That's something, that's very, very strange." And I swear we saw people in there and then just as I said to him, "I swear there's people in there," it just shot off to the left as fast as it could go. I mean it just disappeared. Within a split second, it had gone.

We kept driving and about a kilometer ahead, all of a sudden, there's this really, really bright light in front of us, and I've got my hand up, up above my brow to look out the window, because it's that bright, but I can't see anything.

I said to [my husband], "What are you going to do?" He said, "I'm going to keep driving." From there, that is the last we remembered until . . . I knew I was going to see a UFO, you know, I just knew, because of what we had seen, I'd seen it twice in one night and he had seen it once . . . and the adrenaline is pumping, the heart is thumping, I'm so excited. All of a sudden I'm sitting in the car, and I'm saying to my husband, "What happened?" And he says to me, "I don't know. We must have gone around a corner or something."

By the time we got home he was definite of everything, but at that time he didn't know what happened either. I said to him, "I swear I've had a blackout . . ." because adrenaline just doesn't disappear in a split second like that. I mean your heart is going mad! And all of a sudden . . .

One thing that really annoyed me was that I could smell vomit. I couldn't figure out where the smell of vomit was coming from.

I argue about this half the way home until it started getting ridiculous, and I ended up just shutting up to stop all the fighting that was going to come out of it, you know, because we fight like cats and dogs. . . .

As we were getting close to our home, about twenty minutes away from where we live (there was no one on the road), I saw a figure standing on the side of the road—a tall, dark figure. It was only for a couple of seconds, and I didn't relate it to anything until much later on at all. But it made me turn my head. I kept it in my mind, because it reminded me of a story I was told when I was a little girl about the headless horseman on the side of the road . . . because it was on the side of the road. It wasn't headless or anything, just this tall black figure. I saw it for only a couple of seconds and then I couldn't see it anymore, but I thought I saw it.

Post-Encounter Fallout

I get home. That night I actually had a dream about UFOs to top it all off, that something happened . . . but a whole lot of it went out of my head.

Kelly and her husband argued about what happened for part of the way home. Both agree they saw a UFO, but cannot agree on the feeling Kelly had of experiencing a blackout or missing time and seeing people. They also could both smell vomit and were each experiencing unexplained stomach pain. For Kelly it was like pain from severe muscle fatigue that radiated from her lower abdomen to the upper shoulders.

They arrived home at about 2:30 a.m. Kelly was certain they had left their friends' place before midnight. Her husband refused to acknowledge Kelly's growing concerns and insisted they had left later, more likely around 1 a.m.

Kelly experienced unexplained menstrual bleeding and became extremely ill. She had had her period the week before.

Kelly recalled:

But when I got home that night, that is when I found the triangular mark below my navel, with what I thought was a little laparoscopy cut, and I also started bleeding that night. Three and a half weeks later I ended up in hospital with an infection in the womb. I hadn't stopped bleeding. The doctor said to me, "You must have been pregnant," and I said I was not pregnant. I know I wasn't pregnant. He said that the only way you can get an infection in the womb is if you've been pregnant or it's been caused by an operation. They put me on a drip and fixed me up. I felt pretty crook [an Australian colloquial expression for feeling sick].

[The hospital] actually did a laparoscopy, another laparoscopy. This was not when I first went in. I went back in later, another six weeks after that, because I had a lot of pains in my stomach, and just wanted to have it checked to see what it was. And I still had the triangular mark there. . . .

They just did a blooming laparoscopy cut right next to it. [There was] no comment whatsoever. . . . I have a letter from a friend saying that she saw [the triangular mark].

Conscious Recollections

Kelly's recollections did not emerge from hypnosis. Indeed, she had only one session well after the main investigation of her encounter had been completed, and it failed to reveal anything of significance. Kelly felt she was at best only lightly under and generally thought the session was of little value.

Kelly elaborated on how she came to recall the experience:

We went down to a girlfriend's place a little bit later, a few weeks later, and the subject of UFOs came up and her husband was saying, "Oh, I don't think they really exist." It was my husband that said, "If you had seen what Kelly and I saw you might change your mind!" I said, "What are you talking about?" You know, if I'd seen something, I'd have remembered it. I didn't even remember that I had seen it hovering in the middle of the road. It had been totally blanked out of my mind. And I search my head for days, because I knew he wouldn't say something if he didn't mean it. He was telling me, "Remember on the way home from [your girlfriend's], remember, it wasn't making any noise . . . ?" And I was just sitting there. I couldn't remember it.

And a few days later, all of a sudden I remembered it! It hit me!

And I thought, "Oh." Then I remembered going into the light and then I couldn't remember anything else.

A couple of weeks after that, this started to really bug me, because I remembered that light, and I remember arguing with him all the way home, but it was all I did remember.

I went up to [my girlfriend's] house again in October, this time for her other daughter's . . . birthday, and again we went to Bingo. On the way home from Bingo that night, we went along the same road, and as we passed a certain spot I just got this incredible feeling of terror go through me, I mean absolute terror. All of a sudden I just started remembering, and by the next morning I had remembered just about everything that happened, except there's still missing time that I can't [remember].

What we had actually done, we had driven . . . into the light, but the road curved and the light we had thought was in front of us was actually to our right-hand side. It was in the field, and it was massive. . . . [Estimates put the possible diameter of the UFO at "the size of a house" or perhaps close to 50 meters.] *So it was very big. Why I knew it was very big was because we could have driven for five minutes. The road sort of wound around this part. You could have driven for five minutes and not had it out of your sight the whole time.*

Kelly and her husband had a clear, uninterrupted view of a craft of enormous size. It was much larger than the UFO seen a few minutes earlier and it was at ground level in the field at the bottom of a gully area.

I asked him to stop the car and we both got out. I remembered leaning back in, actually on the floor to pick up my handbag, because I didn't go anywhere without my handbag. And that's one of the sort of things that triggered off a lot of these memories doing that. The other thing was telling myself, "You are conscious. This is real! This is happening! This is real! . . .

I was just listening to myself getting all excited, and I thought I bet-

*ter calm down. . . . I get really excited. For a while it was just ab-
solutely terrifying, but you can't help it, because it sounds really
whacky, I mean this is not the way it's supposed to happen at all. . . .*

*We crossed over the road, we jumped the gutter and we walked
up. . . . I looked down the road and there was another car—a light blue
car—pulled up. Some people got out, and went across the road. I only
thought it was two, but it was actually three, but I didn't pay much attention. They must have been at least a hundred meters down the road
from us. When you've got something like that in front of you, and
you've got people down the road . . . well, I was more interested in
what was in front of me than them, so I didn't get any detail. . . .*

*I'm standing there and we are looking at this thing [for about
thirty seconds]. All of a sudden there is a black figure on the field. It's
about seven foot [sic] tall. . . . I knew it was really tall at the time."*

For Kelly this was quite startling. Somehow she expected
to see a human being, but this was not human. *Its shape was all
wrong.* Kelly tried to use thought as a means to communicate.
She was immediately overwhelmed with fear. Its eyes seemed
to turn to a red fire. At the distance of about 150 meters, they
possessed an extraordinary luminosity.

*It started coming towards us, only slowly, and it had big red eyes.
It sounds stupid, but it had great big round red eyes, like huge fly's
eyes, and they were red like, not like a reflection of red, but like burning red, like . . . fluorescent stop lights, I suppose, that sort of real burning red.*

*All of a sudden I started screaming out [to my husband]. . . . Now
this has really got me baffled because of the fact that a human being
doesn't know this, so I don't even know how I came out with this, but I
started saying, "They've got no souls," and then I started screaming,
"THEY'VE GOT NO SOULS." Then all of a sudden there were heaps
of them in the field, not just one, a whole heap of them, and they
started coming towards us, I mean . . . faster than a man could run,*

and they were gliding off the ground. They got halfway across the field. They split up, some of them went towards the other people [two or three people, Kelly thought] *and some of them* [the rest, Kelly thought] *came towards us.*

Kelly found herself screaming out to the other group of people down the road, *"They're evil! They're going to kill us!"*

The next thing I know I felt this Umph! in my stomach, right across here like I was winded, but I was thrown right back and I was on my back on the ground. I sat up, with my head between my knees. Here, I'm trying to stay conscious. I couldn't see. My eyes. . . . It was all black. I couldn't see anything. I'm screaming out to [my husband], *"I can't see anything! I can't see . . ."*

Kelly speculated that her being "winded" may have been caused by an electric fence at the site. However, this electric fence was inside the main fence line, would probably not have been sufficiently strong to have "winded" her, and most likely was not on at the time. But because of this possible prosaic explanation for her being "winded," she was confused over whether the apparent forces involved were "good" or "evil." She is still uncertain of this and appears open-minded as to what was involved.

But the next thing I heard him saying, "Let go of me." His voice was all sort of cracked up with fear, and I'd never heard that from my husband. He's not frightened or afraid of anything. . . .

Then this male voice said, "We [don't] *mean you any harm." And then he said, "Why did you hit Kelly then?" That's the last I heard of* [my husband]. *No one else talked except me. I heard the male voice. Then I heard myself saying, "Oh, God, I'm going to be sick." I've got my head between my knees, and I just felt like violently nauseous. Then I must have blacked out for a little while. I don't remember being sick. Then I remember hearing talk about being "a peaceful people," and I started screaming out, I said, "Don't believe them, they're going to steal your souls." I know it sounds so ridiculous now, but at the time I was hysterically terrified. I was really, really terrified. I had never felt terror like that—not even in my worst nightmares had I felt terror like that. And it was like . . . oh, there's one thing I remember that he said: "I wouldn't harm her. She's my daughter." Now when I first saw the [UFO] . . . on the way up to [my girlfriend's] in the field, the first thing I did was pray. And I took it as sarcasm straight away. And it sounded like sarcasm. It didn't sound, you know what I mean, it sounded like there was even a small laugh after that. I don't know, it just wasn't good to me."*

Kelly's strong faith in God allowed her to get answers to many of life's situations, albeit sometimes in the most subtle and unlikely ways. For Kelly the brief observation of a possible UFO on the way to her girlfriend's place on August 7, 1993, was perhaps a glimpse at one of life's mysteries—perhaps even a "lesson" from God. So she made a silent prayer, which began with "Father." She thought for a moment, *"Wait for me . . . I'll be back down this way in a few hours."*

Kelly had long been on a spiritual quest, which was anchored in a religious journey and a desire to understand the great mysteries. While she had little time for organized religion, she had a deep interest in the great religious works and the Bible in particular. Therefore, given her brief prayer for clarification of the nature of the UFO event on the way to her friend's on August 7, and the use of the word "Father," she felt "mocked" when during the close encounter of August 8, on the way back from her friend's, she heard a voice saying, *"She is my daughter."* To Kelly there was no way this was her "father," and she said as much before losing consciousness.

Anyway, I started screaming and going on about demons trying to steal people's souls. That's what I did. I know it sounds stupid, but I did. I like not to admit that it came from my mouth, but it did, you know what I mean. But I'm going to tell it the way it is.

Next thing I hear him saying, "Would somebody do something about her." And I felt a hand . . . touch my shoulder. It wasn't hard, it was quite gentle. That's when I absolutely cracked! I'm still sitting on the ground and I couldn't see a thing, but I made sure that my eyes were just fierce, you know what I mean. Something snapped in me. Before that I was crying. All of a sudden something snapped in me and I got so angry. Then I started screaming out, "How dare you do this to these innocent people," like it was my fault. Because I was on a big spiritual search, and I really got the impression that it was my fault. And I thought, why involve other people? . . . I felt like, almost like there was a fight for me. Like it was something I had to do. . . .

Anyway, I started screaming out stupid things; told them to go back where they came from.

All the while Kelly was invoking God, and screaming at them to go back where they came from.

Next thing I remember I was sitting in the car. I've still got missing time.

Her last memory was of driving into the light.

The "Dream" of August 8, 1993

Once home and asleep, Kelly had a bizarre dream. This "dream" is of interest because it emerged within hours of the actual event and it places Kelly back in the encounter. Subsequent "dreamings" do not have this quality about them.

In the August 8 dream, she is on the side of the road with her head between her knees. She becomes aware that she can see again. A being is leading her husband down the slope onto the field.

Throughout the dream she is unable to see the beings

above the level of their elbows. Their limbs seem long and thin.

Somehow she is convinced that the being with her husband is female. She tackles it and then blacks out. In the dream she once again regains consciousness to find herself on the extreme right of the field, with the UFO farther down the field to the left. Before her on the ground is a still body, at first nonhuman, changing to human. A woman, apparently middle-aged, standing farther down the field, was screaming at her, "Murderess, murderess!" She is overcome with grief with no awareness of having killed anyone.

Still in the dream, a hand on her shoulder leads her to follow obediently. Eventually, Kelly becomes aware that she is in a small room, with only a small table and a being standing before her. The being tells her she did not kill anyone, and that they had to use her own sense of morality to overcome her fear. Kelly has a profound sense of knowing this being. On a table behind this being is a Bible, one of hers, which had disappeared a few weeks before. The being offers Kelly a strange choice that heightens her suspicions of the motives involved. She is told she can come but that she would have to leave the Bible behind. In the "dream" the being gives her this Bible.

The "dream" ends at that point. A few days after the encounter her husband found the Bible in the car.

The First "Dreaming"

The first of Kelly's strange "dreamings" occurred before she regained a full recollection of the events of August 8, 1993. It started with a sense of a presence warning her to be calm, followed by a frightening "sucking" sensation; it was as if something was being taken from her. She came out of the dream terrified, only to be confronted by a tall black figure in a floor-length hooded cloak. It was about seven feet tall, with glowing red eyes. She screamed in terror and the being disappeared. To Kelly the being resembled a "soul vampire" or the "grim reaper." The being did not have any UFO context at the time, as her memories had not resurfaced at the time.

In the dream, and it must have only been within a week or two of this event . . . all of a sudden I felt like something was being sucked out of my chest and the pressure . . . and I got scared and it was like a little bit of it came back in. And then it was like a violent sucking again. It's hard to explain because it wasn't . . . I don't know what it is like for energy to leave the body, but it was something like that. And I got that scared in my dream that I woke up and standing beside me and only that far away from me (between five inches to a foot at the most) was a black figure about seven feet tall (because he just reached the top of my mirror) with a hood on it, with big red eyes. I was literally terrified. My husband slept on the couch that night. . . . I woke him up, "Come to bed, come to bed!" But at that stage I didn't remember [the August 8 encounter], so it didn't click.

The Second "Dreaming"

When, after I did remember it, I had another dream and these dreams seemed very physical. I know I'm dreaming and I've got to wake up out of them. . . .

In this particular one, I felt as if my legs were being pulled off the bed, and it was like I was paralyzed from my waist down, and my legs were being pulled over to the side, yet I could almost use the top of my body. Then I'm grabbing a pillow, trying to hit my husband, to wake him up, because he was beside me that night. I'm fighting this. I'm not going to let this thing drag me off the bed by my legs. Then I woke up and saw it standing there again! This time the hood covered the eyes, and it didn't scare me. . . . I was still terrified, but it didn't scare me quite as much, because each time it scared me, it was that same power like I felt out in the field that night.

When I was sitting on the ground it was like something, almost like a frequency or a sound vibration or something. And it's getting right into my head! And I couldn't get it together. Like I was trying to get my logical thoughts together, not logical, almost conscious thoughts, and I was fighting it the whole time; which is probably why I seemed to remember more than my husband or even the other people.

The Third "Dreaming"

The third "dreaming" occurred at a friend's place on October 23, 1993. By then Kelly had experienced the flood of recollections. The two earlier "dreamings" took on an added significance, since the beings seemed to "manifest" after Kelly awoke from them. This time upon waking from a strange "dream" that seemed to take the form of a "peak experience" (in the sense of psychologist Abraham Maslow's term), she saw what appeared to be the same creature, but naked this time and leaning over her, looking as if it were about to kiss her navel. It was tall, with a head much larger than normal, very long and thin arms, and an abdomen out of proportion to its thin frame, like the stomach of a child suffering from malnutrition. Its skin resembled gray-black plasticine. Kelly woke up with a scream. Then from her mouth came "an uninterrupted stream of hysteria." With morning she insisted that her husband check under the car; she had heard a distinct voice during the night urging her do so. He ignored her request and their car subsequently suffered a problem.

Numerous other strange events occurred around Kelly. Some were of a "psychic" nature; others seemed to involve "electrical disturbances." For instance, their car would frequently start up when no one was in it, even though the vehicle had a kill switch that made such events even more startling and curious. These events seemed to come to an end by January 1994, when the fourth and final "dreaming" occurred.

The Fourth "Dreaming"

In the dream, the bathroom light had blown out. Once again there was a sense of a presence. Something was trying to grab Kelly's right hand. Eventually she let her hand be taken. She immediately woke up. Once again the creature in the black robe was standing at her bedside. It went away. It turned out that the light had indeed blown and the diamond and sapphire rings she wore on her right hand had gone. To date, they have not been found.

The Other Witnesses

If the August 8, 1993, encounter had involved only Kelly, it could be argued that the experience may have been some sort of psychological episode. However, the presence of the other witnesses argues that a real event took place. Actually, the investigation carried out by Phenomena Research Australia suggests that the focus of the whole encounter was not Kelly, but two other women who were in a second vehicle.

The PRA's 1993–1994 investigations disclosed that, other than Kelly and her husband, a married couple and their female friend also appear to have witnessed and become involved in the event. Bill, the male witness in the trio, like Kelly's husband, appears to have had only limited involvement. The two women, on the other hand, had a conscious recollection up to and including their onboard encounters. They recollected the UFO and the tall black beings, though their description of the beings did not feature the red eyes Kelly saw.

The three were also in a car and approaching the site of Kelly's encounter when their own experience started. They all heard a strange noise and started to feel ill. Bill felt faint. He started to lose control over the car and ran off the road, striking a pole. After checking for damage, they drove off and within minutes a speeding car with its high beams on passed them. Then another car passed them. They then came to a bridge, which was followed by a sharp turn. Farther along they stopped. Throughout, Bill's vision is impaired. He does not remember seeing the UFO; he seems to have been isolated from the central experience in some unexplained manner. He only recalls smells and sounds and the fact that a lot of activity was taking place. He subsequently underwent hypnosis, which expanded his apparent recollections to being onboard the object as well, but once again these recollections involved smell and hearing only.

The two women with Bill had a conscious recollection of the UFO, the beings, and their onboard experiences. Their description of the UFO matches Kelly's closely. The primary element of their onboard experiences seemed to be a form of examination. They did not visually remember being abducted. Neither of the two women saw each other or any of the others while in the alien environment (ostensibly on board the UFO). They were, however, each somehow aware of what was happening to the other. The women also shared some of the physical symptoms that Kelly Cahill experienced, such as body markings. One of them also had severe ligature indentations on the skin of her ankle.

A third independent witness came forward via an anonymous letter sent to *Who* magazine, a popular weekly Australian publication, during October 1996. The wife of the supposed witness wrote the letter. She indicated that her husband, apparently an employee of the Victorian government Law Department, was reluctant to be identified. However, the letter described some unique and unusual information—specifically a red dragon logo on the spare wheel cover on the back of his vehicle, which one of the women in the second car (not Kelly) had mentioned seeing—that confirmed that the man was indeed present at the scene of the encounter.

This witness also claims to have had ankle ligature marks similar to those reported by one of the women in the second car. The presence of this man's car perhaps also explains why Kelly Cahill was able to notice a second party and their car. It appeared that this third witness had his headlights, which illuminated the second car.

Perhaps for the first time independent witnesses have been able to provide information that enabled researchers by cross-checking to reveal a striking degree of similar information, therefore offering a compelling case for the reality of the strange events described. A range of apparently related physical traces also strengthens the ontological status of the events in the Dandenongs on August 8, 1993, and may represent compelling evidence for a reality behind abduction events. Phenomena Research Australia (PRA) claim they had two laboratories confirm a number of unusual anomalies and magnetic variations at the apparent site of the UFO landing. There appeared to be some interesting changes or differences in soil chemistry—an above-average sulfur content, the presence of pyrene (which occurs in coal tar and is also obtained by the destructive hydrogenation of hard coal), and tannic acid—in a crescent-shaped indentation. Also found at the site was a triangular patch of dead grass on the ground.

Unfortunately, other than a few tantalizing fragments they have doled out, none of these findings have been comprehensively reported by Phenomena Research Australia, despite a passage of what is now ten years. Sadly, even the key witness, Kelly Cahill, who took the brave step of coming forward, has not been made privy to the full data that PRA report they have on the incident. Given my own inability to gain full access to their data on the case, particularly to what seemed provocative physical data, I grew to regret my decision to involve PRA in the first place. As a consequence, a very promising case has not realized its full potential.

But further developments, independent of PRA's own investigations and still not fully elaborated, suggest that perhaps the focus of this extraordinary encounter had not been Kelly or the second (or indeed the third) party on the road that night. Someone else totally unrelated to the parties on the road had apparently also experienced an abduction. In fact, this person's whole family had been subjected to an ongoing alien intrusion. The parties in the cars on the road were, it seems, in the wrong place at the wrong time, innocent bystanders in a much larger alien tableau. The people involved in this hitherto hidden aspect of the Narre Warren North experience have asked that their involvement in this remarkable incident remain undisclosed for the time being.[21]

In this case, possibly the first time, independent witnesses were instrumental in providing compelling physical and other evidence for the reality of UFO abductions. The case could be a striking example of the importance of physical evidence needed to prove the reality of extraordinary UFO events.[22]

Forbidden Science

Over the years I have been fortunate to work with other scientists on this controversial problem in a sort of "invisible college," an informal network of scientists who view the UFO phenomenon as worthy of serious attention. Our work has been done in the spirit of the movement that preceded the formation of the English Royal Society in the early 1660s, when it could be dangerous to be interested in "natural philosophy," or, as we call it today, science. Many of my colleagues in this invisible college prefer to contribute anonymously because work on "the UFO problem" is seen as "forbidden science." What's most frustrating about the situation is not that many scientists are skeptical about UFOs, but that they ignore the compelling evidence that's readily available.

As a scientist myself, I have seen time and time again how rarely high-quality science has been brought to bear on the UFO problem. Too often there is a knee-jerk rejection of even the *possibility* that there might be something to UFOs. The true enquiring spirit of science has not touched the UFO issue. Consequently, popular culture has rushed to fill the void left by the inattention of science.[23]

I have been working with this group of invisible college specialists to bring hard science to the controversial field of alien abductions. The bizarre events that focused on Peter Khoury provided an extraordinary opportunity to assess the reality of abduction experiences. It was time that forensic science confronted the alien abduction controversy. The difficulties and frustrations in gaining full disclosure to the research conducted in Kelly Cahill's case made me more determined than ever to try to extract the full potential of the physical evidence in Peter Khoury's extraordinary abduction encounter.

CHAPTER FIVE

Alien DNA?

T HERE HAVE BEEN MANY IMPRESSIVE CONTRIBUTIONS TO THE relatively new science of astrobiology, the scientific study of extraterrestrial life. *Planetary Dreams: The Quest to Discover Life Beyond Earth,* published in 1999 and written by Robert Shapiro, an expert in DNA research and professor of chemistry at New York University, is an excellent and entertaining contribution to the field.[24] Like most scientists, Shapiro rejects stories of UFOs and alien abductions as credible evidence for extraterrestrial life, although, to his great credit, he at least considers the subject in passing. He rightly highlights the scientific rule, "extraordinary claims require extraordinary confirmation," and argues that "no quantity of personal accounts, photographs of objects in the sky, or sketches of humanoids can ever establish the claim that aliens are here, as too much hangs upon it." Shapiro then of-

fers some suggestions on the sort of evidence "not yet forthcoming, [which] would do quite nicely, however."

Robert Shapiro lists three kinds of real evidence, the third
of which is quite germane to this book:

"1. The recovery of an artifact of a technology well be-
     yond ours. . . .

2. The prediction of a totally unexpected astronomical
   or geological feature, taken from the aliens' maps.

3. The return of an extraterrestrial biological specimen.
   A cutting taken from one of the potted plants that
   the aliens brought from home would serve well
   enough, though a hair or flake of skin from a humanoid would be even better."

While Shapiro presents these examples somewhat tongue
in cheek, he goes on to suggest that the preoccupation of
UFO advocates with extraterrestrials is perhaps misdirected.
He suggests that the "aliens" perhaps come from Earth rather
than "out there," and speculates, "It seems far more likely
that a race of humanoids would have developed as an extra
branch on our family tree on this planet and then sequestered
themselves out of sight to avoid conflict. This would explain
their appearance (somewhat like the science fiction visions of
our own future evolution) and their interest in crossbreeding,
as well as the scarcity of sightings of huge mother ships as
compared to smaller saucers." Thus, a biological sample
would reveal if the "aliens" are DNA-based and related to us,
like the Neanderthals, or if they have an entirely separate origin and thus a totally alien biochemistry. In the end, the skeptical Shapiro concludes that these "UFO aliens" most likely
don't exist and therefore no such biological sample will ever
be acquired.[25]

Shapiro was obviously not aware of Peter Khoury's "alien

hair" sample, which might go some way to respond to his somewhat whimsical speculations. But more importantly, in my view, is that our handling of the analysis of "alien hair" points to the proper path for science to take on this subject rather than its a priori rejection of the UFO field. Peter Khoury's 1992 experience with the two strange women and the unusual hair sample provided us with a remarkable opportunity. But initially my focus was on more conventional investigations, and there was little I could do when I first heard about the sample in 1996.

However, by early 1998 my circle of "invisible college" colleagues had expanded to include some scientists from the biochemical field. They were cautiously interested in UFOs, but mindful of the subject's stature as "forbidden science," they insisted that their involvement would need to be conducted strictly on an anonymous basis. The biochemists were well established in their field, with well-regarded peer-reviewed publications and research. But job security and peer respect are powerful inhibitors of open declarations of interest, particularly given the long and sorry history of science in the UFO controversy.[26]

Because I had an existing and reasonable public profile and strong and established connections with key researchers worldwide, our group, which called itself the Anomaly Physical Evidence Group (APEG),[27] decided that I would be the public face of this biochemical variant of the UFO "invisible college."

And so it was that in 1998, as my new biochemical associates were lamenting the general absence of credible UFO evidence pertinent to their field, specifically biological samples, I pointed out that maybe there was a credible sample that might be worth their attention. I outlined the experience

Peter Khoury had in 1992 and mentioned that a hair sample had been recovered. Rather than recoiling and roundly rejecting the bizarre nature of the story, they were immediately very interested. Our discussions then focused on the fact that techniques like polymerase chain reaction (PCR) amplification and sequencing of mitochondrial DNA on the hair sample might provide a real challenge to the credibility of abduction stories and an opportunity to do some real science in an area that had been dominated by wild claims and theories. I felt that while the prospects for a breakthrough would be very small, the utility and potential of the biochemical and forensic approaches to alien abductions would come as a breath of scientific fresh air.

Could a DNA-based forensic approach help determine the reality of the alien abduction experience? If these bizarre abduction episodes occur at a physical level, at least as we understand it, then a DNA analysis of the controversial hair sample—the "alien hair"—would provide us with a unique opportunity to apply some real science to this controversial area. This would be a rigorous test of the credibility of Peter Khoury's extraordinary claim. I pointed out to Peter that if the testing confirmed a prosaic origin for the sample, then he may have some explaining to do, not only to us but to Vivian, his wife, in particular! Peter's response was that he knew what had happened and was keen to have his evidence tested in this way, regardless of the eventual results.

Phase One—Hair Shaft Analysis—Strange Evidence

The biochemists undertook a standard forensic investigation of the shed hair over several months, from late 1998 to early 1999. The method they chose, after considerable discussion and evaluation, was based on forensic procedures established by the Federal Bureau of Investigation (FBI) for the accurate analysis of biological evidence found at crime scenes.[28] The method was designed to produce a precise DNA sequence to see how it compares with the normal human DNA variations. Research had shown that, for forensic purposes, the greatest apparent DNA variability in humans is found in the region of DNA called "mitochondrial hypervariable region I." Mitochondrial DNA is present in any cell in quantities 200 to 500 times that of nuclear DNA, and hence is more easily detected in an aged sample. It has the further advantages of uniquely preserving the female DNA and being more resistant to deterioration than nuclear DNA. Thus, it is more likely to be recovered at crime scenes, and in our particular situation our "suspects" were female, albeit of a most unusual kind.

The variability or "hypervariability" of a sample provides the basis of the DNA equivalent to fingerprinting—"genetic fingerprinting"—the DNA profile of the source of the biological sample that can uniquely identify its owner. But first the sample must be properly prepared.

To extract fragments of mitochondrial DNA from the blond hair of the strange female Peter Khoury encountered, a two-centimeter piece of hair shaft (located just above the root) was cut from the found hair and transferred to a small sterile tube. Similar pieces of hair shaft were cut from the hair samples that had been taken from Peter and his wife Vivian in my presence and would act as controls.

The three hair samples were easily distinguishable both to the naked eye and under a microscope. The strange blond woman's hair was extremely thin and almost clear. In other words it had very little pigment (melanin) material. In sharp contrast, Peter Khoury's hair was thick and black, while hair from Vivian was of normal thickness and brown. The unusual optical clarity of the strange blond woman's hair made handling a big issue, as it was almost invisible to the naked eye on a glass surface; it could only be handled under reflected light. When the hair was investigated further under a powerful microscope, it was found to be at the lower end of normal human hair for thickness and had a pronounced "mosaic" structure, which was perhaps accentuated by the low levels of melanin.

The hair samples were prepared in the prescribed way for forensic DNA work, which involved an elaborate "chemical washing" (extractions) to ensure that the surface of the hairs was not contaminated in any way. No DNA was recovered from the washing step. Following a prolonged extraction step at elevated temperatures, fragmentation of the hair into smaller pieces, and repeated boiling and freezing with dry ice, the DNA from each was released from its protein structure, purified, and prepared for the PCR amplification step. The PCR step, in simple terms, creates a very large quantity of the originally extracted DNA, making it possible to analyze the DNA in more detail. Without that critical step of amplification and copying, there would not be enough DNA to examine. Special bits of DNA ("primers") carry out this vital step in a kind of large-scale, rapid biological copying process.

The procedure successfully extracted DNA over the hypervariable range of the hair samples from the strange blond woman and Peter Khoury. Vivian's hair sample did not produce PCR amplification products; the analysts concluded that this might have been due to chemical treatment. But as Vivian had not come into direct contact with the blond hair sample, this was not viewed as critical as a good result from Peter's sample.

Concentrating on the amplified products from the blond hair and Peter Khoury's hair, good quality DNA libraries were obtained and then analyzed to see how they compared to the normal human DNA range, namely what is referred to as the "human consensus" region of the vast database of DNA.

The results were striking. As expected, Peter Khoury's DNA lay within the human consensus range, but the blond hair results revealed a fascinating anomaly. Most humans, particularly Caucasian types, would have their DNA falling within this so-called "human consensus" range. Species that are different have what are called "substitutions" in their DNA. These substitutions are "mutations," "transitions," or changes in the DNA that are brought about by the replacement of a basic DNA unit with another. Chimpanzees, for example, differ from humans at fifty-five locations in the same DNA region. The ancient Neanderthal man differs from humans at twenty-seven locations.

The DNA profile from the blond hair sample consistently showed five of these changes, or "substitutions." Three of them seemed fairly common among human racial types, but two appeared to be quite rare and, given the nature of the sample, bizarrely so. In a blond hair sample that apparently came from someone who was biologically close to normal human genetics, there were extremely rare "mutations" that should have been evident only in an unusual racial type—a rare Chinese Mongoloid type, which is one of the rarest human lineages known. The hair usually associated with such peoples is black in color, as would be expected from the Asian type mitochondrial DNA, yet in this sample it occurred in a blond hair sample. The rarity of these DNA mutations is further evident in the fact that they lie further from the human mainstream than any other except for African Pygmies and aborigines.

This was a most unexpected result. Given that the hair was blond to clear in color, and the apparent donor, the strange blond tall fair-skinned female Peter Khoury encountered under very bizarre circumstances back in 1992, we would expect the DNA to match racial types from Scandinavian countries in northern Europe. But the DNA revealed otherwise.

This ground-breaking mitochondrial DNA sequence analysis, utilizing PCR and undertaken by the Anomaly Physical Evidence Group, revealed a strange and unusual DNA sequence that was not the result of laboratory contamination, and that would be found only in a few other people throughout the whole world, but not in blond to fair-colored hair.

The chief biochemist's report concluded: "What implications might these comparisons have, for possible authenticity of the alien hair sample as collected by [Peter Khoury] in Sydney in 1992? While it would not be impossible for him to have had sexual contact with some fair-skinned, nearly-albino female from the Sydney area, such an explanation is ruled out by the DNA evidence, which fits only a Chinese Mongoloid as a donor of the hair. Furthermore, while it might be possible to find a few Chinese in Sydney with the same DNA as seen in just 4% of Taiwanese women, it would not be plausible to find a Chinese woman here with thin, almost clear hair, having the same rare DNA. Finally, that thin blond hair could not plausibly represent a chemically bleached Chinese (including the root), because then its DNA could not easily have been extracted.

"The most probable donor of the hair must therefore be as [Peter Khoury] claims: a tall blond female who does not need much color in her hair or skin, as a form of protection against the sun, perhaps because she does not require it."

Because of these very unusual results, further controlled

DNA testing was undertaken with blood samples obtained from Peter Khoury, his wife, and a male Chinese biochemical analyst. Like Peter's wife, this man had never touched the sample, and indeed had never had any contact with Peter or Vivian Khoury. The mitochondrial DNA results showed that all three new control samples lay very close to a modern human consensus, quite unlike the results obtained for the "alien" hair.

While these phase-one results were not in themselves proof of the hair's alien nature, they were certainly very provocative. How is it that rare Asian Mongoloid DNA was found in a blond hair sample, which, if related to the blond woman Peter Khoury encountered, would have been expected to yield a Caucasian DNA profile? Prosaic explanations such as hair dyeing, an albino source, contamination, and other factors were examined and ruled out.

The presence of the strange Asian-looking woman during Peter Khoury's July 1992 encounter was also considered. Could this individual have somehow introduced her DNA onto the sample? This is not likely, given the external washings of the hair sample and the fact that the results emerged from the total shaft, principally its contents. Peter's case and the DNA analysis of the hair sample were documented [29] and received widespread attention in 1999. But more striking results would follow.

Phase Two—Hair Root Analysis—Very Strange Evidence

The original DNA work was done on the shaft of the hair. More fascinating anomalies were found in the root of the hair when the soft root tissue mitochondrial DNA was extracted. The procedure was essentially the same as the first phase, with the apparent absence of melanin allowing some limited modification that led to the soft root-tissue and hard hair-shaft being extracted separately as independent samples for PCR. The PCR methods remained the same, once again concentrating on the hypervariable region, while one small nuclear region, which contains a special gene (the CCR5 gene) was amplified. Other nuclear regions gave no amplification, apparently because the hair DNA was too degraded.

The new phase-two results on the blond "alien" hair revealed a stunning anomaly. Depending on whether we analyzed the hard hair shaft or the soft root, its mitochondrial DNA appeared to be of two different kinds! From the lower hair shaft we again obtained the same rare Chinese mitochondrial DNA substitution. But from soft root tissue, we obtained a novel Basque/Gaelic type mitochondrial DNA, which had a rare substitution for that racial grouping along with several other characteristic substitutions. This in itself was a stunning result. The testing methodology meant that prosaic explanations such as contamination or laboratory error were ruled out. In any normal human DNA we should get consistent DNA irrespective of where the sample comes from, be it hair, blood, or other tissue. But incredibly here within the one piece of hair we were getting two types of strikingly different DNA, depending on where the mitochondrial DNA testing occurred. The rare Chinese type DNA in the hair shaft was confirmed and a possible rare Basque/Gaelic type DNA was found in the root section. The biochemists had no explanations for this strange anomaly.[30]

The results were very puzzling and controversial until 2000, when a paper appeared in *Nature Biotechnology* that revealed some recent findings in baldness research that used advanced cloning techniques to transplant previously incompatible hair.[31] Here in the sample recovered under controversial circumstances by Peter Khoury back in *1992*, we seemed to have a similar example of combined or "grafted" DNA. Was this evidence of advanced DNA cloning techniques? wondered our principal biochemist? Could the blond female possibly have had a hair replacement in the past?

The only other, though quite remote, alternative is that someone with an extremely rare Asian Mongoloid DNA substitution had forcefully handled the hair sample. Given the circumstance of the recovery of the hair sample, there appear to be only two candidates for such an alternative: the strange blond female or the unusual "Asian" female. This possibility still supports the claim that Peter Khoury was in the presence of two very unusual persons with apparently rare mitochondrial DNA, not easily found in the Sydney, Australia, region. But the results we obtained do not support prosaic processing errors or contamination.

Perhaps even more controversial is that we also have findings suggestive of nuclear DNA indicating possible viral resistance. The hair sample seems to contain two deleted genes for the CCR5 protein and no intact gene for the normal undeleted CCR5. The gene CCR5 is very important in human genetics because it confers immunity to AIDS-HIV when not present. The deleted CCR5 gene factor may also provide resistance to other viruses such as smallpox. While the limited nuclear DNA results were insufficient to achieve the clear result needed, the implications are startling.

The principal biochemist working on the hair sample speculates that, if correct, these CCR5 results hold profound implications for mankind and our place in the universe, specifically "for the nature of our relation with human-like species elsewhere." Both the recent origin of the gene and the fact that both gene copies must be absent to confer viral resistance in humans argues for a "breeding mutation" rather than natural selection. "Could such a double-deletion," he wonders, "have been induced by germline [egg and sperm cells] genetic engineering, as a way of making a whole alien species viral resistant? Alternatively, if delta-CCR5 arose on Earth only recently by random mutation, why did this valuable phenotype [physical characteristic] not show itself first in apes, or in primitive countries such as India or China where smallpox is rife? How could resistance to HIV have been achieved by natural selection 5,000 years ago, if HIV came into man only recently in 1955?

"Today we see delta-CCR5 almost exclusively among people of northeast European descent, as well as Ashkenazi Jews.[32] The origin of that mutation dates to 5,000 years ago, and it is associated with a blond, blue-eyed racial type. Could such a mutation have been introduced into local humans, by cross-breeding with some blond alien species in ancient Ireland or Scandinavia?"

The suggestion of possible viral resistance has some interesting implications for the numerous accounts of "aliens" who seem to interact with humans without any apparent concern over human viruses. Richard Thompson, in his book *Alien Identities: Ancient Insights into Modern UFO Phenomena*, describes an intriguing—and relevant—story from India, concerning "the Smallpox Lady." This unusual woman apparently "appeared at times of smallpox epidemics, and she would mystically cure people of smallpox." This was "a classic celestial woman, as portrayed in South Indian temple sculptures. She had a big forehead, a very thin waist, and very prominent breasts." She seemed to have almost "supernatural powers." Thompson draws parallels between the Vedic female

Devas (or goddesses) and UFO entities, particularly those of the "Nordic" kind.[33]

Together, the two distinct phases of DNA analyses undertaken on the "alien hair" sample recovered from Peter Khoury's bizarre 1992 experience provide a very striking array of genetic findings. They appear to be evidence for advanced DNA techniques and DNA anomalies of the sort we are only now discovering, or starting to make sense of, in mainstream biotechnology.

I certainly would be more comfortable if dozens of samples like that from Peter Khoury's strange 1992 experience existed and that their testing revealed either some consistency or contradiction of results. Either result would provide more coherent answers to this controversy. But for now we have only this anomalous sample, which has provided us with a strange DNA profile. While our findings will continue to be subjected to review and debate, they still represent potent evidence for a forensic and scientific approach to alien abduction cases that will help illuminate the possible strange reality that may be at the heart of such claims.

CHAPTER SIX

The Nobel Laureate

OUR ANALYSIS OF THE ALLEGED ALIEN HAIR SAMPLE WOULD not have been possible without a remarkable breakthrough conceived by Dr. Kary Mullis, an American biochemist. On a Friday night in April 1983, Mullis was driving up to his Anderson Valley cabin in Mendocino County in northern California. During the drive to his cabin, Mullis made one of the great discoveries of modern chemistry—the polymerase chain reaction (PCR), a surprisingly simple method for making unlimited copies of DNA, thereby revolutionizing biochemistry almost overnight. For his discovery, which he described in *Scientific American,*[34] Kary Mullis was awarded the Nobel Prize in Chemistry in 1993.

Oddly enough, this famous and controversial scientist may have had an alien abduction himself. On another Friday night, during the summer of 1985, Kary Mullis drove up to his cabin and arrived at around midnight, after driving for about three hours. Mullis dropped off groceries he had bought on the way, switched on the lights (powered by solar batteries), and headed, with flashlight in hand, to the outside toilet located about fifty feet west of the cabin. On the way, Mullis encountered something extraordinarily weird.

Recalling the episode for his 1998 book *Dancing Naked in the Mind Field,* Mullis wrote, " . . . at the far end of the path, under a fir tree, there was something glowing. I pointed my flashlight at it anyhow. It only made it whiter where the beam landed. It seemed to be a raccoon. I wasn't frightened. Later, I wondered if it could have been a hologram, projected from God knows where.

"The raccoon spoke. 'Good evening, doctor,' it said. I said something back, I don't remember what, probably, 'Hello.'

"The next thing I remember, it was early in the morning. I was walking along a road uphill from my house."

Mullis had no idea how he got there, but he was still dry despite the extensive early morning dew. His flashlight was missing; he has never been able to find it. There were no signs of injury or bruising. The lights of the cabin were still on; the groceries were still on the floor where he had left them. But some six hours had gone by, all unaccounted for. Later in the day he found that an area of his property—"the most beautiful part of my woods"—had inexplicably become a place of dread for him. A year or so later Mullis exorcised this fear John Wayne–style by firing rounds of ammunition into the woods. While the cowboy-style psychotherapy proved successful, it did not help him find out what had happened to him that night in the summer of 1985. His encounter with the talking raccoon resembles other alien abduction accounts where something ordinary appears to be a screen memory for something apparently quite alien. Mullis is probably the only known Nobel Prize laureate to claim an alien abduction–like experience.

Kary Mullis describes himself as "a generalist with a chemical prejudice." Others have described him as "Hunter Thompson meets Stephen Hawking," or "the world's most eccentric and outspoken Nobel Prize–winning scientist." We cannot dismiss Mullis's experience as a drug or alcoholic hallucination; Mullis states he was not under the influence of either that night.

But he is not the only one to have experienced something strange at the cabin. His daughter, Louise, once disappeared for about three hours, after wandering down the same hill. She also reappeared on the same stretch of road. Her frantic fiancé was about to call the local sheriff. Mullis had told no one of his own experience until his daughter called to tell her father about *her* strange experience and advised him to buy a book, Whitley Strieber's *Communion*. By coincidence, Mullis had already been drawn to the book and was just then reading Strieber's report of strange "owls" and little men entering his cabin in Upstate New York.

In his book, Mullis concluded, "I wouldn't try to publish a scientific paper about these things, because I can't do any experiments. I can't make glowing raccoons appear. I can't buy them from a scientific supply house to study. I can't cause myself to be lost again for several hours. But I don't deny what happened. It's what science calls anecdotal, because it only happened in a way that you can't reproduce. But it happened."

Kary Mullis confirmed all this and more when I spoke with him in 1999.[35] He told me then that yet another person had encountered a "glowing raccoon" between his cabin and the toilet. This was a friend of Mullis's, who did not know of the "raccoon" story and was a first-time visitor. The incident occurred during a party at the cabin after the announcement of Mullis's Nobel Prize in 1993. On the way up the hill toward the house the friend encountered a small glowing man, who then suddenly expanded into a full-sized man who said something like, "I'll see you tomorrow." Mullis's friend, who was not experiencing a drug- or alcohol-induced hallucination, left with a friend without informing anyone. They returned to their hotel at a nearby town. That night the man inexplicably found himself outside in the hotel car park, troubled and terrified by the impression that he had somehow been back at the Mullis cabin.

The following night, he and his friend returned to the Mullis cabin, where the celebration continued. As the man arrived he was shocked to see the "full-sized man" he had seen as an expanding apparition the night before drive up in a car. Panicked, he left immediately, holding Mullis somehow responsible for the previous night's events. Sometime later he tearfully revealed the full story to Mullis, who identified the man as his elderly neighbor. Mullis checked with his neighbor, and sure enough, he had come to the party on the second night, arriving to be seen by the terrified visitor. However, he was certain he was not there on the first night, not in person, anyway, and not lurking as a glowing raccoon or a small glowing man that expanded into a vision of himself!

Given this sort of activity on his property, it perhaps isn't surprising that Kary Mullis told me he thinks the nature of his experience is even stranger than abducting aliens. He speculates about multidimensional physics (à la Michio Kaku's *Hyperspace*)[36] at a macrocosmic level: it's "like anything can goddamn happen and the speed of light is not really the limit in terms of interactions with other cultures or whatever. This stuff about grabbing people or subjecting them to all kinds of experiments, it's just anthropology at a level we don't understand quite yet."

When I told him about our PCR testing of biological samples from abductee experiences, he replied: "You might imagine that I thought of that myself, as for instance in 'you can have some of mine, if I can have some of yours.' " He feels that the idea of an alien culture needing our DNA to survive is very unlikely and finds implausible the existence of an alien presence on the scale suggested by some of the bleaker alien abduction tales. Mullis knew of the 1998 book called *The Threat* by historian and abduction researcher David Jacobs, who argues for an alien secret agenda to create alien-human hybrids, which are in the process of integrating into our planet and supplanting mankind.[37] Kary Mullis thought that any culture that could conquer the barrier of space-time could have easily conquered the far simpler problems of complex biochemistry and would not need us in the manner described in the gray alien–human hybrid theories.

The very strange nature of the experiences that befell Kary Mullis, his daughter Louise, and his friend remind us of the bizarre dimension of this alien abduction controversy. Strangeness and reality unbound are the siren calls of this phenomenon. Discontinuities of memories feature aplenty. Extraordinarily bizarre recalls are often ignited by seemingly innocuous events or by revisiting the scene of the encounter. Sometimes memory is seemingly contorted, manipulated, or just shut down during a person's engagement with this strange phenomenon. The boundaries between reality and the unreal become blurred. Maybe Mullis is correct. Maybe our world is intersecting or interacting with some sort of bizarre quantum hyperspatial playground.

CHAPTER SEVEN

Tape of Lost Memory

SINCE I HEARD ABOUT PETER'S 1992 ENCOUNTER MANY months after its occurrence, I wanted to know if what Peter had told other people about this experience was consistent in every detail with what he told me. In trying to track down this material in 2000, I came across some apparently contradictory information he had given in an interview with Moira McGhee and Bryan Dickeson, members of the UFO group Peter had originally joined in order to seek assistance to understand his 1988 experience. The two said they had taped an interview with Peter in which he talked about the strange women, but did not mention anything about the hair. Peter recollected they had taped an interview with him, but thought it was unlikely that he had referred to the strange women, as the interview had occurred prior to his experience with the strange women in the summer of 1992. But even if he had spoken to them about the women, perhaps Peter had been circumspect about the details and not mentioned the hair, given the bizarre nature of the episode and his concerns and issues about lack of support and general acceptance.[38]

To try to clarify this situation, I wrote to both Moira McGhee and Bryan Dickeson, with Peter Khoury's written support, requesting a copy of the tape. It took a few years before Moira was able to find, copy, and give me the tape of their interview with Peter. The tape was not dated, unfortunately, but Moira felt that the interview had probably occurred in either late March or early April of 1992. Did this mean that Peter Khoury and all the other points of confirmation and validation—Vivian, Jamie Leonarder, the diary entries, and other evidence that pointed at a July 23, 1992, date for the encounter with the women—were all wrong? Did it mean that perhaps Peter had fabricated the story and had somehow produced the unusual hair, one that would ultimately reveal an extraordinary DNA profile, which strikingly confirmed the apparent existence of possibly two strange people? The taped interview would, in fact, reveal a much stranger possibility. Could the two women have visited Peter Khoury at least once prior to the July 1992 episode?

For Moira and Bryan the absence of a reference to the hair might have implied it had possibly come from somewhere else. They speculated that perhaps Peter had acquired the hair sample from another abductee, maybe from one of those who came forward during the October 1992 visit of American abduction researcher Budd Hopkins. Hopkins did pass on some Australian abductee contact details to Peter's new support group, but none of these witnesses had experiences that featured physical evidence such as hair samples. Given the prominence Peter's case achieved after its initial documentation in 1999, it seems unlikely that an alternative source for the hair samples would have remained hidden all that time.

Conflicting Perspectives

The idea that Peter might have manufactured both the story about the women and their connection to the strange hair sample is unsubstantiated and simply does not fit the facts. To my knowledge, only one other account of possible alien hair emerged in Australia, in 1992 and 2001. A young Queenslander reported to Peter Khoury in mid-1996 that a strange encounter had occurred some five years earlier with a bizarre-looking creature and provided "hair" samples as evidence. These were quite unlike those obtained in Peter's experience. They were dark in color and looked more like coarse brush bristles than fine humanlike hair. Unlike the scenario considered by McGhee and Dickeson, Peter Khoury did not appropriate these samples and call them his own. In fact, even though he had still not fully revealed his own experience and its related sample, he passed the Queensland sample to John Auchettl of PRA for analysis during a UFO conference in Brisbane in October 1996. Unfortunately, PRA provided few details about their hair analysis, other than saying that the Queensland "hair" sample was interesting, mineral in nature, and had some sort of "memory" behavior, then later saying rather contradictorily that it was of no real interest. My group, the APEG, was eventually able to undertake some DNA examination of samples of this material, but found it yielded only weakly amplified DNA, consistent with human consensus results, which suggested prosaic contamination, namely the kind of results that would come from handling the sample. These results were communicated to Peter Khoury and the source of the sample. No further work was undertaken by APEG, and the young person who originally provided the samples wanted no further involvement.[39]

So we have no coherent evidence that supports a scenario that Peter Khoury fabricated his story of encountering two strange women and recovering a strange hair sample at the same time during July 1992. The hair sample that yielded the extraordinary DNA data clearly came from Khoury's own experience.

Dating the Taped Interview

In order to properly interpret this earlier interview, I needed to nail down the date of this interview with Moira McGhee and Bryan Dickeson. Moira believed the interview had occurred soon after a major traumatic event in the Khoury family—the shocking death of Peter Khoury's sister-in-law, who was murdered on March 17, 1992. She had gone missing for a day and a half, and all her family, Peter included, had been trying to find her. Unknown to most of the family, she had been involved in an affair with a local schoolteacher. When she had gone to meet him to end the affair, he killed her and then killed himself. This tragedy had a devastating effect on the family. Peter was no exception. Moira McGhee recollects that their interview with Peter took place a few weeks later, as she had asked Peter whether he wanted to do the interview at the time because of the family tragedy.

There is yet another element to help date the McGhee/Dickeson interview. Moira remembered that Peter made a point of getting them to look at his head to see where the needle had been inserted during his 1988 experience. McGhee recollected that Peter had to part his hair before she could feel a big lump at the location. This clearly placed their interview as occurring before May 13, 1992, when he received the severe head injury during the job site assault. This injury required several stitches. I have seen a video of his head injury, and McGhee clearly saw only the much smaller lump linked to the 1988 needle insertion and not the much more evident and severe head trauma.

In the wake of that severe head injury, Peter Khoury did experience some brain swelling and short-term memory loss. Vivian recollects vividly having to write many notes for Peter during his long recovery period, to remind him to do things, such as taking his medicine. One has to wonder if his inability to remember a possible earlier episode involving the two women was due either to short-term memory loss or to a deliberate effort by the strange women to block his recall. Or was there no such separate event and are we dealing with a flawed or imperfect memory and forced reconstruction? The severe head injury and its short-term consequences also do not provide an explanation for the whole July 1992 event, as we have ultimately the presence of the hair to explain and the extraordinary DNA results that in a remarkable way confirm aspects of Peter Khoury's strange encounter.

Some internal evidence on the tape adds further support to an interview date occurring between March 17 and May 13, 1992, indeed before late April—around Peter's birthday. Moira McGhee asked Peter Khoury, "How old did you say you were now?" Peter's reply: *I'll be 28 in April.* Because of the poor quality of the taped interview, it is uncertain whether this statement means "I'll be 28 this coming April," or whether it's a rhetorical statement that means "I'm 28 in April" (coming or passed?).

In the interview Peter suggests that this incident with two women in the bedroom of his home occurred between four to six months previously, making it either October or December of 1991! During the interview he asks his wife Vivian to join them, saying: *You know when I said to you I had some weird feeling that there were two women in the bedroom? How long ago did that happen?* [Peter then says in response to feedback from Vivian] *About four months. . . . About six months ago.* There follows a brief discussion mentioning two jobs Vivian had either in late 1991 or early 1992. These served to prompt her confirmation that the "incident" Peter was asking her about occurred about six months before the interview. As this dating exercise seems to suggest that the interview probably occurred in late March (or, less likely, early April), then an event apparently involving the two strange women mostly likely took place in October 1991.

A Different Encounter?

In the taped interview Peter Khoury talks about what apparently happened in 1991. In a very broad sense there is some similarity between the events described and those that occurred in July 1992. Two strange women were involved, one Asian looking and the other blond European looking. Both encounters occurred in the morning and took place in the bedroom of the home, and in both cases Peter seemingly bit one of the women on the breast. But beyond these broad similarities, there are a number of quite significant differences.

In the event Peter describes on the McGhee/Dickeson tape, the women's roles are reversed. It is the Asian looking woman who straddles him, while the blonde takes a passive, apparently observational role. The "Asian" is naked, but the blonde is described as wearing "overalls," even though Peter also refers to her as having pale skin "all over." Peter describes the duration of the 1991 event as lasting ten to twenty minutes, considerably longer than in his recollection of the July 1992 event. Peter then wakes up two hours after the encounter; in the July 1992 event he calls Vivian at work just after she arrives there, which is less than an hour after the incident occurred. In the 1991 event Peter describes much more activity with the Asian woman. She takes Peter's hand to her breast, taking it back and forth three or four times, then some eight to ten times she cups Peter's head to her breast, back and forth, seemingly wanting him to bite her breast area. There is no mention of any hair being recovered in this taped interview regarding this apparent 1991 event.

When I first heard this taped interview, I wondered if the differences between Peter's two recollections were the result of flawed recollections of only one event or, less likely, support for a fabricated story. But it soon became very clear to me that Peter seemed to be talking about an entirely different episode than the 1992 event, one that apparently occurred about nine months earlier.

Confronting a Dilemma

I contacted Peter Khoury after I had listened to the tape and advised him that in the interview he clearly refers to the two strange women but doesn't mention the hair sample. I also indicated that there were significant differences between the two apparent events, but I did not describe these differences. Peter and Vivian were thoroughly puzzled and confused by this news. They couldn't understand how Peter could be talking about the two strange women before the date of his head injury. As often as Peter turned the events over in his mind, the only thing he could recollect with certainty was what had happened in 1992. He could not recollect an experience nine months previously.

Peter was willing to try anything to sort out this apparent confusion, even suggesting hypnosis, truth serum, or a lie detector test. I had serious reservations about the utility of hypnosis as a means of accurately retrieving memories, but after discussions with some consultants we concluded that before Peter Khoury listened to this long lost interview tape, it might be useful to see if hypnosis could shed any light on the situation. As the episode had occurred in a rather stressful year for Peter, between the murder of his sister-in-law and the head injury from the job site assault, my friend and consultant hynotherapist Robb Tilley thought that these events could be used as "signposts" for hypnotic recall. Dick Warburton, a friend and psychologist, consented to monitor the sessions, which were carried out and videotaped on November 6, 2003.

This would not be Peter's first experience with hypnosis. When he joined his first UFO group, he came into contact with a hypnotherapist, who indulged in New Age techniques and asked leading questions. The session was very unsatisfying for Peter, who did not achieve a trance state. Peter later revisited his 1988 experience under hypnosis with a much more experienced hypnotherapist, Frankh Wilkes. A recollection of an alien presence emerged, but it provided little clarification beyond his conscious recollections. This regression occurred just two days before the July 1992 experience with the two strange women that led to the recovery of the hair sample. The hypnotic "reliving" of a previously consciously recollected experience and the close proximity of another strange incident made me wonder if there was a causative connection. But later a case for the manipulation of memory and reality by factors other than hypnosis would emerge.

When Harvard psychiatrist John Mack visited Australia to research the subject of alien abduction in 1996,[40] he took the opportunity to regress Peter to try to clarify the 1988 experience beyond the moment of blankness after the needle was inserted in his head. Under Mack's regression Peter described being taken into a room that was lit up. The walls were white, as if the surface was the light source. He was on a table, with one entity standing above him and speaking to him with a sound like the chirping of birds. It was one person, a shadowy tall figure, but it sounded like there were fifty of them. Peter was thinking at the time, How am I going to remember what you are telling me? The scene just faded and everything went dark again. That was all that came out in the session. Ultimately Peter was not really satisfied with the hypnotic recollections and feels more comfortable with the consciously recollected details.

The 2003 Hypnosis Session

Following the induction by Robb Tilley, Peter Khoury eventually entered a deep state of hypnosis. During the first phase he seemed to have had some sort of restraint or containment in his wrist or arm area, but beyond verbal expressions of discomfort it was far from clear if this was related to any particular incident. Robb Tilley brought Peter out of this situation. It was clear that the session was not going to focus spontaneously on any episode with the women unless he was directed to it. After the next induction, Robb Tilley gave Peter a specific direction:

RT: You are very sick. . . . We want you to see yourself in the car with Vivian. Were you driving her there?

PK: *Yeah.*

RT: Okay. You are driving off towards the station.

Peter elaborated on what Vivian was wearing. It is clear she was dressed up for a cold day. This seems consistent with the July 1992 dating (winter), and not a date around October or November (early spring). It was clear with the direction about feeling very sick that Peter would elaborate on the July recollections. We hoped that by starting there he might later refer to another episode.

PK: *It's really cold.*

RT: What time is it?

PK: *It's seven. Five past seven. . . . I've got to get her to the station before the train leaves. . . . I pull over and get sick, just vomiting.*

RT: What does Vivian say?

PK: *I think she is used to it by now. She doesn't say much about it.*

RT: So you drop her at the station. She gets out of the car and then what do you do?

PK: *She's telling me that I may need to see a doctor, just to watch myself basically. I'm on my way back. I pull over twice. On the way there I pull over about five times. I get home a lot quicker coming back. Can't stop . . . God. Couldn't stop the car quick enough.*

RT: So you're back at home.

PK: *Yeah, I'm back at home. I sit on the bed. I actually get a tissue box. I put it next to me. I've got it next to me, because I was worried I would vomit in the bed.*

RT: Are you cold?

PK: *No, I'm pretty rugged up.*

BC: What are you wearing, Peter?

PK: *Winter clothes, long pants, long-sleeve jumper, blue in color, a tracksuit, but not matching. . . . But the sloppy Joe is a lot newer . . .*

BC: So you lie down back on the bed.

PK: *I'm in bed covered, trying to get to sleep.*

BC: What time was that, when you go back to bed?

PK: *About ten past seven, quarter past seven at the latest. I'm toss-ing and turning, trying to relax.*

BC: So you've gone back to sleep.

PK: *Yeah.*

BC: Something woke you at some point?

PK: *Yeah. . . .*

BC: Do you remember anything waking you up?

PK: *Someone's jumped on the bed, not jumped on the bed, more like patted the bed, slightly tapping it.*

BC: Is that what woke you up?

PK: *Yeah, no, that's what woke me up, so pressure on the bed, nothing . . . it's not like rocking me out of my sleep, it's just something light, like flattening the sheets, throw your hand over and tap it, but very light movement.*

BC: So, what's going on?

PK: *I'm just trying to find out what this movement is? . . . I'm sit-ting up.*

BC: Have you opened your eyes yet?

PK: *No . . . no, not yet . . . I've got my eyes open. I can see two fe-males in the room.*

BC: Do you want to describe them? Where are they?

PK: *I'm sitting up now. There is one directly in front of me at the end of the bed and another one at the corner of the bed, at the foot of the bed almost.*

BC: Describe the one directly in front of you.

PK: *I'm just trying to catch up.*

BC: What's happening there?

PK: *It's like there's two of me. It's like ah . . . it's like a see-through, ah, how can I put this . . . it's like a balloon image, a clear balloon, that's shaped like me. It's a clear image of me. I know it's me and I can see her through that, through this image. . . . [long pause] I'm catching up with my body . . . it's confusing . . . just as my physical body catches up . . . I can hear the Asian-looking one . . . saying things about babies, babies, more babies, something about babies, not children . . . something about babies, another baby . . . this is really strange. . . .*

BC: How are you hearing that?

PK: *Just in my head, telepathically, I suppose. She's looking at me . . .*

BC: Is she telling you this?

PK: *Yeah. Yeah . . .*

BC: Are you seeing her lips moving?

PK: *No, no, but I know she's the one telling me, because we are looking at each other, and the other one's just sitting there. She's almost straddling me, on top of me.*

BC: This other one that's sitting there, almost on top of you, can you describe her?

PK: *Very attractive, very, very attractive.*

BC: What's happening now?

PK: *She's reaching out to me, like cuddling me.*

BC: How do you mean?

PK: *Like a mother's cuddle in a way. There's nothing sexual about it, just cupping . . . hugging my head, and hugging me towards her breast or chest. She's got my face in her breast and I can't breath.*

BC: What are you feeling at that point?

PK: *Just shocked. I'm just shocked by it. Definitely shouldn't be in my bedroom, that's for sure, but . . . how they got here I haven't got a clue. But . . . it's definitely got a sexual aspect to it, because I mean they're naked in front of me. There's no sexual act as such.*

BC: They are both naked, are they?

PK: *Yes. Yes.*

BC: You were saying the Asian woman was talking about babies.

PK: *Yes, as soon as I caught up with my body, it's as if I could communicate with them, where before there was that detachment I think and I couldn't relate. Once I caught up and I became one . . . it was strange, it was like she was already talking, but as I became one I can hear her now.*

BC: What were you hearing?

PK: *Just something about babies, babies, and this new baby will join other babies. They're not children as such, like they're not teenagers, but babies, and that this one will join the other babies. I'm getting a feeling like they're my babies? Like she's telling me this, because it concerns me. She's either touching her stomach, or rubbing her stomach, maybe patting her stomach. It's definitely something about babies, that this one will join the other babies.*

BC: How long does that communication go on for?

PK: *Seconds . . . five seconds . . . not even . . . It's just that I'm hearing her saying babies, this one will join other babies. I feel like my babies. Viv and I haven't had children yet, can't be my babies.*

BC: So is that all she talked to you about?

PK: *Yes, that's all I can hear.*

BC: So what about the woman in front of you, you were saying she had taken hold of you, like cuddling you.

PK: *Yes . . . There's nothing sexual to it, it's just like she's hugging me, but my face is in her breast. I'm having difficulty breathing. I push away from her. She doesn't let me go. She's still got me. From the second she's touched me she hasn't let me go.*

BC: So where is she holding you?

PK: *She's cupping my head, she's hugging my head, the back of my head like with both of her hands. . . . She hugs me again. I'm trying to push away. . . . I can't push away. My face is just buried.*

BC: Why can't you push away?

PK: *She's just strong. I'm trying to push, it's not like . . . it's not like I'm not . . . She's just got me . . . she's in control. . . . Whoo . . . I've got to move, I've got to move.*

BC: Why do you have to move?

PK: *Because I can't breathe.*

BC: Why can't you move?

PK: *Oh, my face is in her breast! It's like, God, it's trying to breast-feed or something! It's just burying my face into her. . . .*

BC: So how long does this go on for?

PK: *Not long at all. Maybe fifteen seconds. . . . I'm just trying to get a breath.*

BC: So what happened?

PK: *. . . Some of her breast is in my mouth.*

BC: What part of her breast?

PK: *Part of her nipple area. . . . Oh gee, I'm just going to push away.*

BC: So you're pushing away.

PK: *I can't push away. I don't know what I've got to do. I'm going to bite. I just took a little nip. Damn, oh gee. She's pushed me away.*

BC: She's pushed you away?

PK: *Oh yeah.*

BC: So what's happening now?

PK: *I just bit her on the breast!*

BC: What's she doing?

PK: *Just pushed me away and looking at me. Oh . . . She's looking at the other one now. They're talking.*

BC: What are they talking about?

PK: *It's not words but . . . they are talking telepathically. I can sense, I can feel it. It's like a feeling.*

BC: So what are you feeling?

PK: *That I've done this wrong. It should have been like the last time. It's different to the last time? It's not the same as the last time.*

BC: What do you mean by the last time?

PK: *Another time. It's like I've got the impression she's saying, this didn't work out, it didn't work out like the other time.*

BC: Who is giving you that feeling?

PK: *They are both looking at each other. I really can't tell which one's saying it. I don't know, maybe the blond one because I bit on her, but I'm definitely getting this thing about, it's done wrong and the last time somehow went better, and I've done something wrong. Things have gone wrong and they didn't pan out like before. . . . I've got this thing stuck in my throat. . . . God . . . it's just like eating something and having a little piece of food that's stuck in your throat. It's a little bit irritating. Whoa . . . it's starting to burn my throat.*

BC: What's starting to burn your throat?

PK: *Whatever I bit, I think. It's just starting to feel really . . . Whoa . . .*

BC: So how is it burning, could you explain that?

PK: *It's strange, I never felt like this, felt this thing before. . . . It's more stinging, I think . . . Yeah, it's stinging in my throat.*

BC: Whereabouts in your throat?

PK: *Deep, it's not like I could spit it out if I wanted to. It feels down near my Adam's apple, a fair way down. . . . Geez, it's really there. It's irritating. I've started coughing. Oh, gee, it's really uncomfortable. . . .*

BC: What about, can you see the women still?

PK: *. . . Not after I'm finished coughing. No, after I've finished coughing I look up and there's no one there.*

BC: How long were you looking down for?

PK: *Oh . . . five seconds. Just coughing, probably five seconds, and they're gone.*

BC: Did you see them go?

PK: *No, just gone, there's just nothing in the room, no one there, and I'm on top of the bed, I can see my clothes . . .*

BC: What about the dooner [a bed covering usually filled with duck down or feathers and used during the winter, which occurs in June and July in Australia]?

PK: *I'm lying on top of it. It's like the bed has been made. I'm just lying on top. It's not messy. It looks too well made actually for me to have been sleeping in it. It's too perfect.*

BC: What about the thing in your throat now?

PK: *It's really irritating. I'm just going to eat something or drink something.*

BC: So what do you eat or what do you drink?

PK: *Just going into the kitchen. I'm looking around the house like . . . I'm expecting anything to happen. . . . Strange, like I know someone's in there. There isn't anyone in there.*

BC: Why do you think that?

PK: *I don't know, and I didn't see them go.*

BC: So did you see them again?

PK: *No. I just feel that someone's there, some sort of energy is there.*

BC: That . . .

PK: *It's like something's there. . . . It's just like I'm getting cold chills, like somebody walking over your grave, or something like that feeling. I don't know, because I didn't see them leave, I'm thinking they're there. . . . I'm in the kitchen. I'm having something to eat.*

BC: What do you eat?

PK: *Just bread, just Lebanese bread, dry.*

BC: Do you have anything to drink?

PK: *Not yet, no, just even the bread trying to make this thing go down my throat.*

BC: Does that make it feel better?

PK: *No.*

BC: So what's happening? What's it feel like at the moment?

PK: *It's just stinging, like I want to cough, but I know if I cough I'll never stop. . . . I drink water.*

BC: So how much water do you drink?

PK: *Easily half a small bottle . . . the little water bottles, about 500 milliliters.*

BC: So how does that go down?

PK: *Not a problem, but the irritation's still there, the scratchy feeling is still there.*

BC: So it's a scratchy feeling now?

PK: *Yeah, stingy, scratchy, makes you want to cough. . . .*

BC: So is that burning feeling still there?

PK: *Yeah, burning, yeah. . . . I'm sitting on the lounge in the lounge room.*

BC: Anybody else there?

PK: *No. No, I'm just sitting down, drinking more water.*

BC: So how long do you drink the water for?

PK: *I drink the whole bottle, about two minutes. . . .*

BC: So after you've drunk that whole bottle, how does your [throat] feel?

PK: *Throat's sore still, just irritable.*

BC: What about the burning feeling, any better or any worse?

PK: *No, the same, just the irritation, you just wish you could scratch it. . . . I got to go to the toilet. I got to do a piss.*

BC: So did you go to the toilet?

PK: *Yeah, yeah, I'm just trying to do a piss, just waiting. . . . There's a lot of pain on my penis . . . oh gee, like I've scratched it, I don't . . . I start to piss, it hurts, but, oh, not on the inside. . . . It's like I've cut my penis. I think I've scratched it. . . . I pull my skin back. I see a hair. It's cutting into my skin. Gee, that looks like it's been there.*

BC: What's that hair look like?

PK: *It's curled, like an "S," but more curves than an "S," just like figure 8's but not joined, but it's cutting into my foreskin, no, not my foreskin, my penis, because I've pulled my foreskin back and it's sitting on my penis, and there's two of them. One's a short blackish, darker than the other one, and one's really like . . . frail like a baby's hair almost . . . frail looking, almost looks like a . . . like that string you can sew with, but it's nylon, and it's not fishing line, really thin, it doesn't even seem real.*

BC: What about the darker one?

PK: *The darker hair just looks thicker, darker, thicker, shorter. The long hair, I've taken the short one off, but not the long one yet.*

BC: So is the short one easy to take off?

PK: *Oh, it hurt, but it's only a short piece. I'm just going to take the bigger one off. That hurts a heck of a lot more but . . . I don't really want to peel it off.*

BC: Why's that?

PK: *Because it's sore, I'm worried it's even going to scar me. It's like there's a cut on my penis. I'm just peeling it off slowly, really, really slowly, so I'm not ripping it off.*

BC: So can you see whether it's cut you?

PK: *Yeah, it's like it's . . . like embedded. I wouldn't say cut as in bleeding, like it's left a mark there, like a groove, the shape of how it was sitting. Got it all off.*

BC: You still have both pieces?

PK: *I'm holding both pieces in my hand. I'm holding them in my fingers, just looking at them. I'm actually thinking whether I should dump them in the toilet and flush them, or keep them.*

BC: What do you do?

PK: *I'm walking out of the bathroom, back towards the kitchen. I get a plastic bag from my briefcase. . . . I've got small different size bags, sample bags. My penis is still sore. . . . I've got four different size bags, big ones, smaller ones, and I pick the smallest one and I stick both hairs in it.*

BC: So what do you do with the bag?

PK: *It's a resealable bag, and I've sealed it. . . . Yeah, I'm seeing my old office here, which is next to the spare room. I go to the filing cabinet in the corner. Actually before I do that I get a sticky paper, a white paper, and I write on it . . . "hair sample" and I've taped it over the bag. The bag's sealed and I've sealed it over again, with the sticky paper. . . . It's in my filing cabinet. I don't go back to bed. I go back to the lounge. I still feel like there's someone or something's in the house . . . something not finished or . . . or they're not ready to leave, but I can't see anyone.*

BC: So how long does that kind of feeling last for?

PK: *It's probably about quarter to eight, eight o'clock, I know that Vivian gets to work at about eight. I want to call her, but I can't, because she's not there. So I'm just going to wait. I want to talk to Vivian. I'm waiting.*

BC: Have you looked at the clock?

PK: *No.*

BC: How do you know the time?

PK: *Just from how long things have taken. We didn't have a clock in the lounge room. . . .*

BC: So do you still feel there's someone else still in the house?

PK: *Yeah, there's still the same feeling, it's like half an hour after all this started I'd say maybe. Twenty minutes, half hour. I can't see anyone.[41] . . . I've got this bad coughing, this really annoying coughing.*

BC: What about after this time, does your penis still hurt?

PK: *No. I can't feel the pain anymore, but my throat's so irritable.*

Peter explained that he tried to spit out the irritant and was fighting for breath, because he was coughing so much, but he was unable to get anything out. The irritation was so bad, it felt as if barbwire were being dragged through his throat. He drank a lot of cough mixture in an attempt to alleviate the irritation, but it brought about only a slight improvement. By about 8 a.m. the feeling that there was someone else there receded. He thought of calling Vivian, and he eventually did so about 8:20 a.m., telling her in between a lot of coughing that something strange had happened. Peter said in the session that he showed her the hair that evening, and told her that he had swallowed something and was coughing all the time, and that the hair was involved, but he didn't tell her about the women, indicating instead that he wasn't ready to tell her. Vivian told him just to tell her when he was ready.

While we were still on the subject of this experience, I returned to Peter's impression that the encounter had seemingly not gone as it should. The feeling he got was that it hadn't gone like the previous time, but he couldn't recollect a previous time or another occasion on which he had met the women. However, he had the impression that they were conveying to each other the idea that they had done this before, and why did it go wrong this time? Why didn't it go like the last time? Peter could only remember this incident. He was certain that it happened after his head injury, as he strongly recollected feeling sick and the vile smell of his medication.

Peter added that when the Asian woman was talking about babies, he got the impression that "this one will join the others," that he was creating something, that it was going to be taken up there, and that he would be able to interact at a later date. She patted her stomach, and at the same time either pointed at the blond woman or to the sky. He seemed to think it was more likely that she was pointing at the other woman but toward her face. Even under hypnosis he was unable to recall another time or another event that involved these two women.

Through the Mirror More Darkly

Robb Tilley took over the session again and asked Peter to go forward to the next strange event, as we had not been able to elicit memories of any other event involving the women before the 1992 episode. What Peter began talking about was an experience in 1996 in which he described being taken through the bedroom mirror.[42] Peter had already told me about this consciously recollected episode, but it turned out that in the regression he went further than his conscious memory—beyond the mirror, giving a striking description of that transition through it, beyond and ultimately back again.

He was in bed with Vivian and he began to feel a sense of creeping paralysis that was increasing in intensity. He tried to touch Vivian in order to break the paralysis he was feeling, but he couldn't.

PK: *I can see the mirrors in my room, a mirrored wardrobe. . . . There's a whole group of beings coming through it. . . . Fuck, it's an army of them. . . . Gee, they are quick. It's like they are levitating or hovering. They are like the little hooded guys. They are only small. . . . There are a few of them. I can see three of them near my bed. One stands at the end of the bed. There's one on Vivian's side, and there's one on my side. . . .*

Oh shit, oh . . . I'm just getting lifted off the bed . . . no one is touching me. It's like a cushion. I'm just slightly off the bed. . . . I'm being floated out towards the mirror . . . back the way they came. There is only three of them. I don't know what happened to the others. . . . The one at the end of the bed has gone through the mirror. He has just faded. Oh wow . . . ohh . . . just vanished. . . . I'm starting to go through it. My feet are going through, oh, it's like I'm jumping in jelly . . . but there's like static electricity, whoo . . . on the outside of my body, like on my skin and my hair. My whole body's like it's got an electric cur-

rent . . . it feels thick. . . . I can see the mirror, one on either side of me . . . they are not touching me. They are off the ground, hovering, levitating, just floating. We started off really quickly, like floating quickly towards the mirror. Now it's really slow. I'm just going . . . It's up to my hips. [It feels] really thick. Fuck, how am I going to breathe through that. . . . Whoo . . . Up to my chest . . . Ohh . . . I'm looking at my face. I can see my reflection in the mirror. They're already gone through. The others have gone through. I've just seen the other two go through. Jeez, I'm up to my chin . . . Ohh . . . I can see my face . . . Ohhh . . . Oh fuck, I'm either scared, or I'm just in total shock, I've got this . . . I'm going through it. . . . It's just really thick . . . it made like it's cotton, sheep's wool, I don't know, it's more like jelly. . . .

BC: What happens now?

PK: *I'm definitely not in my room. . . . I know I'm moving . . . I can see like light patterns, different shades of light. . . . I don't know where I am. . . .*

BC: What can you see?

PK: *There's no one around me. They are all gone. . . . I'm still traveling through. It's like a tube. . . . I've stopped. . . . It's just a big room, huge, but nothing in it. . . . I'm lying on something . . . it's like a conveyer belt. . . . It's like I've been traveling on this. . . . I can feel my body resting on something. . . . It's like I'm sitting on a cloud . . . like you're floating. . . . My mouth's really dry. . . .*

Peter felt like he was there for a while, an hour or two, apparently alone "floating" stationary in this strange room. He tried to relax, to rest, as it is draining to fight the continual paralysis that he is feeling. Then it finally begins to move again. The room is narrowing down. He then starts to feel a lot of static again all around his body, a tingling sensation that seems to be making him cold, and making his hair stand on end. It seems to be becoming very cold, and then he seems to think he isn't wearing any clothes because of the sensation of coldness all over his body. After what seems to him to be a period of about three minutes, it feels like everything is reversing and he is going back. Everything he experienced before now seems to be playing out again, but in reverse. He then finds himself back in his bed, without feeling himself going through the mirror. He now has a very bad headache and he can't open his eyes because of the pain. Eventually he can see that Vivian is beside him in bed asleep. He slowly wakes her up and they discuss what happened. She wasn't aware of anything unusual happening, apparently having slept through it.

Robert Tilley tried to trawl back and forth in time again, but little of value emerged. Despite a number of efforts we were not able to get any clarification of an earlier experience with the two women. Peter Khoury's account of the July 1992 experience under hypnosis was largely consistent with his conscious recollections. There are some elaborations of detail and some elements that are new, but perhaps these were just subconscious reflections of his struggle to try to recollect more fully what had happened. After all, he had come to the session primed with the idea that a tape existed that revealed details described by him of an earlier encounter with the two strange women. Robb Tilley, Dick Warburton, and I were all in agreement that Peter wasn't trying to make these elements up, but there was some debate about what his description represented—fragmented and distorted aspects of one experience or evidence of at least two separate experiences.

The "Forgotten" Revelations

It was now time to let Peter listen to the interview he did with Moira McGhee and Bryan Dickeson back in 1992. The audio quality of the taped interview is poor, with frequent breaks, talking over each other, and the segments are difficult to decipher.

PK: *... in the morning about 7:30 ... I dropped her off at the station at Regent's Park. I came back home.*

MM (Moira McGhee): To here, this house?

PK: *Yes, back here, and I went into the bedroom, lied down in bed. I didn't have work that day, so I thought I'd sleep in.*

These words immediately suggest that Peter is not referring to the 1992 experience. There is no reference to the fact that in July 1992 Peter wasn't working, and hadn't been for some time because of the illness that followed the severe head injury he had received a few months earlier. Clearly an entirely different incident is being described here.

I lay down. It might have been half an hour later ... when all of a sudden I saw myself sitting in bed, sitting up in bed.

There is no reference to what prompted Peter to wake up, unlike the apparent later episode, where he sensed something get on the bed.

MM: You could see yourself?

PK: *Yes ...*

MM: What angle of view did you have?

PK: *Like I was watching myself sitting there and I'm sitting there, basically next to myself. ...*

The apparent later episode features a similar perspective of seeing a transparent version of himself, but from immediately behind, not from beside himself.

MM: Beside yourself.

PK: *Yes ... There were two women. One had straight black hair, like Asian. ... Her face was like leather. ... She looked like she was about thirty. ... Her skin was incredible. ... It was like it was tanned ... tanned a lot darker* [like from under a tanning lamp]. ...

The blond one had really nice goldy blond hair and it was really high, it was like curly wavy . . . and it was really way out . . . and it came down to about halfway down her back, but really puffed up. It was a lot of hair. She was white. Her whole body was white. It was milk white.

MM: How old would she be?

PK: *She looked like she'd be about twenty. She looked a lot younger than the other one.*

MM: What were they wearing?

PK: *Actually the one in the white was wearing like overalls, but the one with black hair wasn't wearing anything.*

Moira's question about clothing leads Peter to say that the blonde was in overalls and the Asian was naked, although earlier he had said that "her whole body was white." He described both women as being naked in the 1992 encounter.

The impression I got was the blond-headed one was watching and learning, and the other one was either teaching her, or I'll explain it to you and you'll see what I mean.

The activities of the two women are reversed in this episode relative to the 1992 episode.

Now the black-haired one grabbed my hand and brought it to her breast. She'd want me to touch her.

The touching of the hand to the breast was not described at all in the July 1992 account.

The other one was sitting on the bed, was just watching. I could see the look on her face. I thought, Why isn't she doing anything?

MM: Were their facial features more like us or different?

PK: *They were a lot like us, but like I said, her face was leathery. One to me looked really young, and her skin hadn't been in the sun too long.*

MM: What were the nose and mouth like?

PK: *Small nose, very, very small nose . . . same features as us . . . I saw lips, but I didn't see teeth, I didn't see a tongue. . . . The one that was really white I couldn't see any eyebrows, but even the black one I couldn't see eyebrows. Maybe I didn't notice them. . . . I didn't notice the color* [of the eyes].

In the 1992 account, Peter does refer to the eye color. A discussion follows about the limited visibility of eyebrows, then Peter continues.

PK: *She grabbed my hand and she'd bring it to her breast. She did that three or four times. . . . Then she grabbed my head and brought it to her breast and . . . This was incredible.*

MM: You were watching this happening?

PK: *I could see this happening.*

MM: Were you fighting it?

PK: *No, no, no. I was just under control. . . . She grabbed my head and I'd do it. I had no control over it. . . .*

In the July 1992 account Peter describes himself as progressively trying to resist this activity.

Now she grabbed my head. She'd bring it to her breast. I didn't know what they wanted to do. The impression I got was that she wanted me to, like, bite on her nipples and I did at one point, and when I'd bit she'd push my head like she'd want pain in a way, and at one time, this happened for about eight to ten times, where she like grabbed my head, put it to her breast, then she'd push me away.

This reaction of the Asian woman in the 1991 event is contrary to the reaction of the blond woman during the more fully described episode of July 1992. Rather than seemingly seeking pain, she reacts and conveys a sense that this is wrong. Peter describes his head being pulled to the Asian's breast as happening eight to ten times in this 1991 episode; in the July 1992 episode he alleges that it only happened three times and on that occasion it was the blonde, not the Asian.

In the 1991 episode, the Asian pushes him away, and in July 1992 he pushes the blonde away.

Then at one time, it was like I could see, honestly, like I could see her breast stretch. That's why I know her chest skin was like leather skin, was like leather. I could see it stretch like that, and really stretch. That's why I got the impression that she wanted me to do it that way because she'd push, not like I'm hurting anything, but she'd push me to, and I could see her breast extend, stretch out.

BD (Bryan Dickeson): Sort of like leather or like rubber?

PK: *It would be more like rubber because it was stretching. . . .*

There seems to be a break in the recording of the tape. Apparently missing is a description of Peter swallowing the piece of skin and starting to cough.

MM: . . . did it stop almost as suddenly as it started?

PK: *See, I wasn't coughing unintentionally . . . I was coughing because I needed to get it out. . . .*

MM: . . . did the coughing ease up over a few days until it gradually diminished or did it sort of stop . . . ?

PK: *Because I was coughing on purpose, when it dissolved or when it just went I didn't like think about it much . . .*

MM: So the cough didn't go on . . .

PK: *. . . no, no, once that thing was in my throat, it was like something sitting in your throat in a particular spot, and you try to cough it out, and once it wasn't there I wasn't coughing. I didn't need to. That's what it was. I wasn't like coughing like . . .*

BD: I need to ask, were you under any stress at the time?

PK: *No . . .*

This is further support for an earlier episode. Peter and Vivian clearly recollected the July 1992 event as happening when he was ill from the head injuries, and because he was not working for some time it was a period of some stress. A discussion follows about whether Peter felt okay about being regressed, and it seems he wanted to deal with both this experience and his 1988 experience.

PK: *It worries me that it might be worse, like I might have actually had intercourse and things like that, but I don't remember it, but then it really sort of affected me, like this was a turnoff, so I can imagine how bad that experience would have been. . . .*

MM: Bit of a personal question, but when you woke up, you must have stopped to think was there any evidence to you that you had had intercourse.

PK: *Yeah, I did, no, nothing.*

MM: There was no evidence to you that you had?

PK: *No . . . as far as I was concerned I still had my clothes on. I couldn't see, no I couldn't.*

MM: No sign? . . . No feeling that you had either?

PK: *No.*

This is a clear difference between the two encounters. In the July 1992 event, Peter does have evidence that there was something on his penis, namely the hair that was recovered.

BD: Was there any feeling of lost time?

PK: *Not that I could tell, but I mean for the short time the experience took to the time I woke up it worked the opposite. Where usually you have missing time, like you're traveling, say, from reports I've read you're traveling, all of a sudden you see a light or something . . . ten minutes later you go home and you find out it's hours gone by. This was different. The experience itself took, say, ten or twenty minutes, but when I woke up it was between 9:00 and 9:30. It was two hours actually.*

This event seems much longer than the encounter of July 1992. Also, in the later experience Peter apparently doesn't go back to sleep, and rings Vivian just after 8:00 a.m. There is no sense of two hours of missing time in that encounter.

So it worked the opposite. You know what I mean?

There was some confusion about what Peter meant here. Some discussion followed, but then the tape pauses and starts on another unrelated matter.

Elsewhere in the interview, Peter first mentioned that this incident with the two women in the bedroom of his home occurred between four and six months previously, making it either October or December of 1991. Given that this is nine months before the more certain event in July 1992, in which Peter in recent fragmented "recollections" and under hypnosis recollected the Asian woman implying a pregnancy or children, one wonders if the two events were separated by a gestation period familiar to us all here on Earth.

Reconciling and Rationalizing?

When Peter listened to this taped interview, he was shocked and surprised. And he was frustrated that he could not remember any of the specific elements of the encounter with the women described in the interview. But he has no doubt that the episode described was different from the 1992 episode. He said, "It's a whole different scenario, very similar, but different things."

That night I played the whole interview for Peter and Vivian Khoury together, particularly in an effort to see if it triggered any recollections in either, but also to see if any aspect could shed light on the whole dilemma of the tape. Vivian was also puzzled by Peter's account of the episode with the women. While most of it was disturbingly confusing and puzzling, she did recollect the detail of Peter's stretching the skin of the woman's breast. As for the rest, she, like Peter, did not remember that account of the women. Both agreed that it seemed to describe a different encounter, and all the available information indicated that the incident preceded the July 1992 episode, and appears to have taken place late in 1991. The evidence for two separate encounters with strange women is compelling. When this interpretation is embraced, the nature of the interactions becomes clearer, and in a way a kind of alien logic or behavior gets laid out for our contemplation.

If we have clear evidence of two separate encounters with these strange women some nine months apart, then the seemingly inexplicable behavior and details described originally in the July 1992 episode almost, in a bizarre sort of way, make sense. The incident from about October 1991 has Peter at home, as he is not working that day. In contrast to the July event, Peter is not ill. Two women appear in his bedroom under inexplicable circumstances. The naked Asian-looking woman at first grabs Peter's hand and takes it to her breast three or four times, then she grabs Peter's head and pushes it back and forth to her breast eight to ten times, seemingly at times trying to get him to bite her nipple. It almost seems to be some sort of bizarre and highly awkward attempt at foreplay. Apparently Peter bites her nipple and seems to swallow a bit of it. During this activity the blond woman, dressed in some kind of overalls, seems to be watching and learning. When Peter begins coughing, the women take their leave, we are not sure how. Peter goes back to sleep or wakes up after about two hours. There seems to be a definite sense of missing time. Some nine months later—after a gestation period, perhaps, for the Asian woman—the two are back for a return engagement. This time it's the blond woman who brings Peter's head to her breast. But it ends far more quickly this time. When Peter bites her breast, she reacts, indicating to the Asian woman that it hasn't gone like the last time. The Asian may be giving a sign that she is (or was) pregnant—she has a somewhat distended stomach. In the beginning Peter was under the bedcover. At the end he was on top of it. There is some sort of unnoticed discontinuity in the flow of events. While Peter is coughing, the pair disappears again, but this time a hair that might have an exotic link to both of them is left in a very obvious place—on his penis. He remains awake this time. There is important evidence of the women's presence to be recovered and a story to be told.

Without this physical evidence it would be tempting to attribute the 1992 encounter to perhaps the medication Peter was on. While not generally linked with inducing major sensory disturbances such as hallucinations, some may have had these factors as possible less common side effects in isolated cases. However, the hair evidence, which supported rather than conflicted with the reality of the encounter, was hardly the stuff of hallucinations. During his extended period of medication Peter only had this strange experience once and the apparent 1991 encounter occurred without medication, injury or stress as possible factors.

But why did Peter not remember the 1991 episode? Why Vivian didn't is perhaps understandable. While supportive of Peter, her priorities during this difficult time revolved around family, not aliens. Since then, her only reference point was Peter's clear, consistent and often repeated recollections of the 1992 episode and the hair recovery. Could the job site assault have caused memory loss, or were other factors at work? Could his encounter in July 1992 have been instrumental in preventing recall of the earlier experience? We may never know for certain, but we do know that the July 1992 encounter occurred. The DNA forensic work has given us an extraordinary level of certainty of that event. All the evidence argues that the July 1992 experience and the "hair of the alien" are not about a hoax, a delusion, or some fantasy.

CHAPTER EIGHT

More Hair

ASK ANYONE WHAT AN ALIEN LOOKS LIKE, AND JUST ABOUT everyone will describe a small, bald gray being with large dark eyes—an iconic description that dominates modern society's perceptions of alien life-forms irrespective of one's belief in the reality of their existence.[43] The cover painting of the alien "visitor" on Whitley Strieber's 1987 book *Communion* served to entrench this image widely in the public consciousness.

The extraordinary public dissemination of the alien image via the cover of this book initiated a massive tide of reports. One of these reports Whitley Strieber described as "the best description of the gray beings" among the thousands of letters that were sent to him in the wake of the success of *Communion*. According to Strieber this was the only case to emerge that featured a key detail that he told me he had suppressed from his descriptions of the "visitors." He did this as a way of trying to evaluate the reports he expected to receive. I am not sure Whitley Strieber was suggesting that all his "visitors" had this detail, or that it was only evident in some. But for him this one letter was a striking personal validation. He asked me to investigate the experience detailed in the letter, since it had come from Australia.

The key element that Whitley Strieber told me he had left out in his descriptions of his "visitors" was the presence of hair. He had omitted this element from his book in order to have a way of later identifying possibly authentic descriptions of the "visitors." The fact that this largely hidden aspect of the gray alien image appears to have emerged in only one of the thousands of letters Strieber received may be a potent argument for some critics that the majority of letter writers might have just been jumping onto the alien bandwagon and reporting only the bald alien depicted on the cover of *Communion*. Such an argument is found wanting, because my point here is that the presence of hair in reports of aliens is in fact quite widespread. Peter Khoury's 1988 experience featured beings that were very similar to Strieber's literary rendering in *Communion*, but Peter became aware of Strieber's encounters only after his own experience. Peter's account of the July 1992 episode featuring the rather humanized "alien" women offered up in a very bizarre way the key feature that had been largely hidden or ignored by many researchers: hair. This alien characterization, a key feature of this abduction event, emerges as an extraordinary "lightning rod" for a new perspective on the case for alien reality.

While it seems reasonable to argue that the bald gray alien is a dominant type of entity in reports of abductions, to argue that it is the whole picture is to impose a very skewed perception on the issue. Embracing the idea that aliens might feature many human characteristics, such as hair, propels the quest for validating alien reality into a very real and not so alien dynamic: the biological realm. Suddenly every biological trace or contact can potentially be transformed into a rich source of data to be calibrated against the huge human genetic database. The very sense of the alien that precluded easy human embrace now becomes that little bit less formidable a task. Now it is a task that can be done and is being done, with verifiable results. If some of these alien contacts involve at least partial human traits, then suddenly the chances of validating that reality are much improved.

Visitors in New South Wales

The case that Strieber passed on to me took place in late 1976 in New South Wales, Australia. It occurred in the middle of the day. A woman in her early thirties was vacuuming her home. Suddenly she felt sick and faint, and decided to sit down on her lounge chair to steady herself. She noticed three strange figures in front of her—"three little beings," she called them.

One being, which the woman understood to be female, was about five feet tall. It was very slender and had a very elongated face. The other two, which she instinctively understood to be male, were much shorter. They had broader, chubby faces. All three figures had large eyes, and the slightest hint of what appeared to be noses and mouths. Each wore a shroudlike cloak, the taller figure in black, the two smaller ones in brown. All had wispy bits of hair at the backs of their heads. In the case of the female, the hair was black; the others' hair was brown.

The taller figure began to communicate with the woman, not verbally, but apparently through the mind, telling her that she had to go with them. The figure continued to insist that the woman go with them, but the woman refused. Eventually the "female" figure indicated the woman could leave. The woman got up from the couch and began to crawl toward the front door. This proved to be very difficult, as her movements were very slow. When the woman reached the front door, the beings ostensibly pulled her back "with their minds." The beings continued to tell the woman that she had to go with them. Again she refused. Once again they "told" her she could go. Again she crawled up the hallway. This time her husband appeared to be there. The woman clung to him desperately. She had a strong recollection of the smell of the pure wool sweater he was wearing. The beings however, "just tore" her out of his arms and pulled her back to the couch again "by force of mind." They said she would go with them and it was useless to fight.

The next thing the woman remembered was hearing her husbands's car pulling up out front. When he came into the house, the visitors were gone. Eventually she was reassured that it really was her husband this time. By then it was 5:30 p.m. More than five hours had passed, of which she could recollect only the first hour. The woman was most disturbed by the thought that her two young children were outside in the backyard playing throughout all of this. They were unaware of what had happened. Her husband did not witness anything except for the woman's profound fear.

Years later, the woman was advised by her sister to read Whitley Strieber's book *Communion* because of the similarity of the events described in it to her own experiences years earlier. The woman never completely read the book, but wrote to

Strieber about her experience. On February 20, 1988, Whitley Strieber called me to give me some details of this and some other Australian events. He said he had asked the witness to write to me. I spoke to the woman on the phone after receiving her letter and later interviewed her extensively in person when she came to Sydney. A qualified clinical psychologist subsequently hypnotized her because she wanted to improve her recollection of what the "beings" looked like. The session revealed little beyond the conscious recollection, only some slight clarifications of the beings' appearance, a brief impression of being outside, "floating feelings," the act of looking out a window, the feeling of strong pressure on her forehead, an impression of some kind of object "like a bolt, with a nut, attached to a cord," plus some other fragmentary recollections. When she started to feel uncomfortable describing the experience, the session was ended. She found that focusing on the beings' appearance frightened her and she did not want to continue. The psychologist and I respected her wishes, and no further sessions were undertaken. For the woman, the most strikingly similar image to that of the beings is the Wandjina—the haunting Australian aboriginal rock paintings that reveal figures with large dark eyes and no mouth.[44]

When Whitley Strieber came to Australia in October 1988 to promote the paperback release of Communion, he asked to meet with the witness. After mentioning the case in Transformation, the sequel to Communion, as "the best description of the gray beings" he had received,[45] he again referred to her experience in his third book about the visitors, Breakthrough, at a time when he was struggling with the possible "negative" nature of the "visitor experience." He described this case as one of the most compelling "negative" letters he had received, and he suggests that the noise of the woman's vacuum cleaner may have somehow prevented the disabling "sonic attack" often used to render people unconscious, thereby allowing her a clear and conscious awareness of the beings' presence. Strieber was deeply disturbed by "the bullying, the compulsion, the cruel and extremely clever deceits" of the "visitors" in this case. Strieber believes her resistance led to the negative aspects of the experience. While Whitley Strieber has reconciled this aspect in his own mind, there are many who cannot embrace it, and just as many, perhaps, who have no need to, because their experiences are more positive.[46] It is a puzzling and frustrating issue with no certain answers at this point, while we still grapple with the question of the reality of these experiences. In *The Communion Letters* Strieber once again addressed this case: "The fact that the incident unfolded in the middle of the day suggests that the visitors are much freer to operate in our world than we have realized. Indeed, their imperious conduct in this case might indicate that they place themselves as a higher authority, with rights over us similar to those we claim over the lower orders of earthly creation."[47]

While most descriptions of alien beings present in abduction episodes, particularly the so-called "grays," refer to hairless beings, the presence of alien "hair" is noted in a significant number of cases. Descriptions of tall "Nordic"–type beings have been reported frequently, perhaps most notably in the Travis Walton case of 1975. Not only did Walton report an initial encounter with the small fetus-like aliens, he also claims to have seen four tall human-looking people, three men and a woman, each very similar in appearance, with the "same coarse, brownish-blonde hair, with the woman's longer, past her shoulders." There are many similar cases in the literature. For example, in her books, UFO researcher Jenny Randles describes a 1976 case from Bolton, Lancashire, in England. A young woman was the subject of an apparent abduction. Subsequent regression described a tall blond female entity that looked human. This entity was over six feet tall, with hair "so blond it was almost white." [48] Had the alien hair been recovered in these cases, a DNA analysis could substantiate whether these accounts were something more than just bizarre stories.

While reports of "grays" have been by default the focus of most UFO abduction research, a closer scrutiny of the history and reality of abduction experiences argues powerfully that the issue is not all about the grays. The preoccupation with reports of grays may have misdirected a generation of researchers and perhaps delayed a more realistic appreciation of what might actually be going on.

The Alien Reality

It's worth noting that the biological dimension of the alien abduction mystery has been present from the very beginning of the controversy. Most researchers in this field recognize that there are actually two "first" abduction reports that have had a strong impact on our perceptions of the phenomenon. [49] These are the watershed case of Betty and Barney Hill from 1961 and the controversial sexual "seduction" encounter of Antonio Villas-Boas from 1957.

During the night of September 19, 1961, Betty and Barney Hill, a respected Portsmouth, New Hampshire, interracial couple, experienced a strange "interrupted journey" on the way home from a holiday. They claimed to have had a prolonged UFO sighting, during which they saw figures at the windows around the UFO. After the experience, they became aware of a period of missing time during which they were unable to remember what had happened. They arrived home a couple of hours later than anticipated.

The couple began experiencing problems—anxiety dreams for Betty and stress-related problems for Barney—that became so severe that they sought the help of a Boston psychiatrist, Dr. Benjamin Simon, in January 1964. Dr. Simon soon learned that the 1961 "interrupted journey" seemed to be the focus of the couple's anxiety. He used hypnosis to explore the experience, hoping to rectify their problems. What emerged was a complex account of abduction and medical examination by aliens on board a flying saucer. Betty Hill said that during the experience the aliens inserted a needle in her navel—a "pregnancy test."

Dr. Simon never believed that the Hills' UFO abduction was real. For him, their story was a fantasy to be explored in an effort to relieve the associated anxiety. Of the "alien's pregnancy test" Dr. Simon said, "There's no such medical procedure that would use a direct needle through the abdomen to examine a condition of pregnancy. This is the sort of thing that makes me doubt the story of the abduction. It simply has no basis in medical practice." Since the publication of a best-selling book about the case, *The Interrupted Journey,* by John Fuller in 1966, a number of medical techniques similar to this "pregnancy test" have been developed, in particular the amniocentesis test and the use of the laparoscope, which is used for both internal viewing and for ova sampling of "test tube" babies. This reference to the "alien pregnancy test" as early as 1961 and development of a similar technique after the publication of the book on the Hills' encounter is at least provocative. The widespread exposure of the book did not immediately unleash a flood of similar "interrupted journey" or abduction accounts.[50]

Betty Hill's "anxiety dreams" were the first documented accounts of the possible abduction dimensions of the Hills' "interrupted journey." She wrote down the details of her nightmares during November 1961. They were largely consistent with the recollections that emerged under hypnosis, but there was a significant difference in the description of the strange beings they encountered: "Their complexions were of a gray tone; like a gray paint with a black base; their lips were of a bluish tint. Hair and eyes were very dark, possibly black."[51] They also had prominent noses. Under hypnosis the aliens described did not have hair or big noses. So from the beginning there was reference to hair, albeit via traumatic dreams that preceded the hypnotic recollections that took place in 1964.

The Antonio Villas-Boas experience occurred on the family farm located near São Francisco de Salles, in the state of Minas Gerais, Brazil, on October 15, 1957. Villas-Boas claimed that while plowing a field at about one o'clock in the morning, he was abducted by three small beings and taken aboard a UFO, where he was forcibly undressed. A clear, thick liquid was spread over his skin. A blood sample was taken. He became ill and even vomited. The most unbelievable aspect of Villas-Boas's story followed. He claimed that a naked woman then entered the room. She had white skin and an unusually shaped face, which seemed to come to a pronounced point at her chin. Villas-Boas stated, "Her hair was fair, almost white (like hair bleached with peroxide), smooth, not very abundant. . . . She has big blue eyes, rather longer than round, for they slanted outwards. . . . The contour of her face was different . . . she had very high prominent cheekbones. . . . Her face narrowed to a peak. . . . [Her body] was slim, and her breasts stood up high and well-separated. . . . She was much shorter than I am. . . . Her skin was white and, on the arms, was covered with freckles . . . her hair in the armpits and in another place [pubic area] was very red, almost the color of blood . . ." Sexual intercourse followed, which Villas-Boas attributed to the effects of the liquid that had been applied on him. He was to comment in an interview with Dr. Olavo Fontes, professor of medicine at Brazil's National School of Medicine, that "some of the growls that came from her at certain times nearly spoiled everything, as they gave me the disagreeable impression of lying with an animal."[52]

There seems to be some striking similarities between the strange blond-haired woman who coupled with Villas-Boas and Peter Khoury's blond interloper, though differences are also apparent. While Peter didn't recall seeing pubic and underarm hair on the woman, DNA profiles that emerged from the shaft and root of the hair yielded the rare Asian Mongoloid DNA (usually associated with dark hair, but found in this blond hair) and the rare Basque or Gaelic DNA (usually associated with orange or dark hair color, but again revealed in the blond hair).

Hair color, in biochemical terms, is determined by varying levels of types of melanin—eumelanin and pheomelanins. The former dictates dark hair and the latter controls brown hair. Natural blond hair has a rather murkier origin, some arguing it is dominated by eumelanin levels or at a genetic level via MC1R (melanocortin 1 receptor) gene diversity. Given that the Villas-Boas blonde was said to have freckled skin and blood-red pubic and underarm hair, it should be noted this is usually associated with high levels of pheomelanin and low levels of eumelanin, linked at a genetic level by nonfunctioning alleles in the MC1R gene.[53] While hair and skin color are becoming increasingly better documented in terms of their biochemical and genetic origins, these issues have still not been fully resolved. While some success has occurred with the elaboration of the genetic nature of red hair, the rest of the story is more difficult. The British journal *New Scientist* reported in July 2002: "Hair color is usually determined by the cumulative effect of several genes, so there's no such thing as a single gene for blonde hair that could be turned into a simple test, for example."[54] The field of genetics may offer UFO research important genetic information that may prove significant in revealing the reality behind the bizarre stories of alien abduction.

More Sex with Extraterrestrials

Peter Khoury and Antonio Villas-Boas are not alone in their strange experiences. There are a surprising number of similar sexual "alien" cases. Many come from South America. A 56 year-old Chilean man called "Gaspar MH" claims to have had a sexual encounter in 1978 with a short woman with white-pinkish skin and large blue eyes.[55] Another case from Brazil in 1978 seems rather too similar to the Villas-Boas case, namely the alleged sexual encounter of Jose Ignacio Alvaro with a tall "silvery"-haired woman.[56]

Then there is the case of João Valerio da Silva of Botucatu, Brazil, who apparently lost consciousness as an attractive naked brown-haired and dark-skinned woman touched him during his 1982 abduction. João Valerio was subjected to a follow-up forensic medical examination by Dr. Luciano Stancka e Silva, who revealed that João Valerio had a circular mark on his breast with unusual lesions within it, and slight lesions on his penis—both elements that recall Peter Khoury's 1992 experience and that of Credo Mutwa in the late 1950s. A strange "oil" was also found on his body and shirt. The shirt was handed over to São Paulo police for analysis, but they later reported that it was lost.[57]

But few of these cases inspire any confidence. Take, for example, Alvin Guerra's alleged "rendezvous with a Nordic nymph." This Texan from Fort Worth reported a very friendly abduction into a UFO, seemingly manned by an all "Nordic" women. He described being "rewarded" for his cooperation with a sexual encounter with the tall, blue-eyed, blond-haired "Nordic" commander. After round two of their prolonged sexual liaison Guerra lost consciousness then woke up naked in his bed, with his penis "chapped, irritated, and in some places raw." This story is one of a number described by C. L. Turnage in her unabashed book *Sexual Encounters with Extraterrestrials.* She claims to have interviewed the man herself, but does not provide any information or evidence to inspire any confidence in the truth of the story.[58]

Turnage also relates the story of a lusty, green-skinned alien lover with long red hair, named Quetyal, who conducts regular home visits with "Susan," a young rural Texan waitress, until his sad demise in a "saucer" crash! Again, Turnage provides no information that could verify this improbable tale.[59] Turnage lists twenty-one cases in all, her first being "Antonio Villa Boaz" (she means Antonio Villas-Boas) and the last "Jesus—the hybrid messiah."

Eve Longren's book *The Love Bite: Alien Interference in Human Relationships,* picks up another bizarre sexual thread. For Longren the "love bite" is "a kind of 'psychic rape' whereby the victim is abducted, then manipulated into bonding with a targeted love partner chosen by the alien beings."[60] This seemingly dubious theme is echoed in cases presented by more mainstream abduction researchers, however.[61]

Genetic Agenda

Whatever their validity, the claims of sexual encounters with aliens are cited by many researchers as support for the idea that alien abductors have some form of genetic agenda. But this notion, among all the fantastic claims made about abductions, is one of the major stumbling blocks to the credibility of abduction claims. Mainstream science argues that if alien life exists, it is unlikely to be compatible with human life. Hence, any claims to the contrary, such as alien abductions, are regarded as absurd.

Therefore, the focused DNA profiling technique we used in the Peter Khoury case goes to the very heart of one of the key claims behind the alleged alien abduction agenda. It provides an opportunity for testing the credibility of such claims. If such claims are true, then there should be some compatibility in the DNA of alleged alien specimens, but some possible anomalies may be evident that would perhaps not be readily reconciled with our measures of human DNA variability. And indeed our original analysis confirmed the "alien hair" in the Peter Khoury case came from someone who was genetically close to normal humans, but of a very unusual racial type.

Perhaps Peter Khoury's two females are the "hybrids" that are said to be the goal of the aliens' genetic agenda. But the whole concept of hybrids in abduction accounts was until recently difficult to reconcile with our understanding of the limitations of interspecies breeding. Indeed, given the possibility that we may be dealing with a vastly technologically superior species that could be biologically different from us, hybrids of aliens and humans would seem scientifically improbable and logically implausible. On the other hand, if the aliens have mastered space travel or interdimensional travel, it seems they would have already conquered the biochemical barriers that normally bar interspecies breeding.[62]

But the pace of development in the fields of genetics and biochemistry in the last decade has extended the horizons of this debate substantially. Indeed, transgenics, the transfer of foreign genetic material into other genomes,[63] perhaps addresses the logical interspecies barrier argument against human-alien hybrid claims. Leading abduction researcher Budd Hopkins and his co-author Carol Rainey highlight transgenics, or rather an alien variation on it, as the breakthrough—"the breakdown of the barrier between species"—that possibly legitimizes the idea that alien abduction may really be about genetic experimentation by alien beings. Their book *Sight Unseen* also focuses on strange accounts of human or humanlike entities they label as "transgenic beings" that seem to have stunted emotional ranges and carry out unusual social activities, sometimes with a sexual element to them (seen also, it seems, in Peter Khoury's encounters).

One case that Hopkins and Rainey describe is that of "Sally," a young woman they view as "highly credible." Sally had been in contact with Budd Hopkins since 1987, when she was trying to understand a peculiar "missing time" experience. Her story featured encounters with the ubiquitous gray aliens, but she also was plagued since childhood by the attentions of a strange man she called "Stewart." Under hypnosis Sally described a harrowing memory of being raped by this man at the age of six. Stewart became a regular specter in her life. He seemed human in appearance—"a tall man [in his middle thirties] with a long, thin face, curly blond hair, and oddly shaped eyeglasses"[64] highlighting "piercing, electric blue eyes."[65]

While those details hardly make Stewart stand out, his persistence and behavior do. He was a frequent participant in her alien abduction experiences, often seemingly taking a facilitator role, a process that usually led to his giving her over to the gray aliens. His attentions on a few occasions were those of "a more violent sexual abuser." Over a twenty-five-year period he appeared remarkably ageless, and "seemed gifted in the paranormal sphere, apparently being able to pass through closed doors or windows alien-fashion, able to communicate telepathically, and able to impose paralysis on human abductees and possible witnesses."[66] Such claims invite dismissal of Sally as being perhaps delusional; however, Stewart may have been more than just the stuff of alienated fantasy.

During the early 1990s Sally was working in Washington, D.C., and sharing an apartment with "Hannah," another young woman. Stewart turned up one night, somehow appearing by Sally's bed, even though the apartment was on an upper floor and the windows and door were locked. Although frightened, she joined him in the living room, sharing the couch, drinks, and detailed discussion about the mundane aspects of her life and job. At one point she questioned Stewart about his reality and nature, but drew only smiles and avoidance. Determined to get an answer to her question, Sally, having noticed that his shirt was opened a bit, "suddenly reached inside, took hold of a long, curly chest hair, and pulled it out. He winced and gave her an angry look, but she was pleased to realize that on some physical level he was real and not a phantasm. A few moments later three small gray aliens approached and she was taken out the window and into a hovering UFO for a more typical abduction experience."[67]

Budd Hopkins points out that Sally phoned him within a week about her recollections of that strange nocturnal interlude. Two things convinced Sally that her experience was real. In the morning she found the two glasses of half-finished drinks that she had served for herself and Stewart. More provocatively, though, Hannah, her roommate, had woken up during the night with "a roaring sound" in her head. Frightened and unable to move, she heard voices coming from the living room—Sally's voice and a man's voice. Still unable to move, Hannah must have gone back to sleep. Hopkins interviewed Hannah by phone, confirming these details.

A few years later, "Molly," an abductee from Budd Hopkins's New York support group, stayed with Sally in another apartment she had taken in the Washington area. Sharing the bedroom one night, they both awoke to find themselves paralyzed and the room bathed in light. Comparing their recollections the next morning, they each recalled a shared abduction experience. Both recalled that several small alien beings had appeared in the bedroom, and both mentioned the presence of "a tall, curly-haired blond man with oddly shaped glasses." Sally knew who it was—Stewart. Molly saw the man "operating inside the UFO along with the small gray aliens." Sally asked Molly, an artist, to draw the man she saw. For Sally, Molly's drawing was an even better likeness of Stewart than her own drawing.[68]

While it is understandable that the presence of a used glass and a hair did not leap to the center of the investigator's inquiry at the time, today's new DNA paradigm promises to verify the alienness of any such stories in the future.

Another "Alien Hair" Sample?

It should be noted that the "alien hair" sample obtained in the Peter Khoury case is not the only one we subjected to DNA analysis. There was another alleged "alien hair" sample, but it did not show the kind of anomalies we obtained in Peter's case. A local businessman, whom I'll refer to as "Mike Wood," from the mid-north coast of New South Wales near Foster, had initially contacted the Australian UFO Research Network's telephone hotline to report some unusual experiences.[69] What seemed to be unfolding was a slowly escalating UFO and alien abduction dynamic that also seemed to involve the witness's daughter. Members of the group UFO Research (NSW) had been in touch with Mike and on the weekend of April 13–14, 2002, they stopped by to visit him. According to Mike, something very strange had happened and he had something to show them. He gave them a small sample bag containing a hair sample, one that he said had come from an alien female. Because of that development, and the incident's apparent similarity to Peter Khoury's 1992 experience, the group contacted me and eventually gave me the sample for possible DNA testing. Before doing so, however, I wanted to establish the credibility and context of the story myself with the witness on location.

I discovered that Mike was not the only one in the area to have seen UFOs; so did others in the community. He had also had some unusual entity encounters. Some of the episodes went back to his childhood. He viewed the phenomenon as something dark, malignant, and ultimately manipulative, and seemingly intensely interactive. He cited one UFO sighting while fishing with his son-in-law. Mike warned him to be careful what he said as he had found that often things he said would be bizarrely reflected back in some sort of UFO display.

As if on cue to the son-in-law's request to see purple coloring in the UFO, the UFO shot up the coast, returned three minutes later, and seemed to blow out to a 200-foot radiant display, with a bright purple center. Then it turned off like a light switch. "I told you," said Mike to his son-in-law.

If that element of a highly reactive, reflected phenomenon was valid, then it may lend some credibility to the incident that Mike claims led to the recovery of the new hair sample. During the evening of Sunday, April 7, 2002, he and his partner were having a bit of an argument. In a somewhat cynical and flippant way Mike said, "I wish one of these space women would come and take me away." That said, eventually they retired to bed. The details of the incident that follow are somewhat explicit.

Mike recollected fragmentary images of a woman's face and her washboard stomach. He remembered her eyes, which were slanted and dark colored. He sensed that there was another woman as well, who seemed darker skinned. There was an overwhelming feeling, almost like electricity going through his body, that powered an erotic dimension to the experience. He woke up still feeling hard in his member: "That night I had the hardest erection I ever had in my life." He did not think he had an orgasm. There was no smell, and no feeling in his scrotum.

When he woke up he remembered having what seemed to be this erotic dream and his raging erection. He became aware that he had a fine white hair wrapped around his penis. He started to pull it off and then dropped it on the bathroom floor. He began to have flashes of memory of encountering two strange women during the night, one blond-haired and olive skinned, the other apparently Asian in appearance. He then remembered the hair, returned to the bathroom, picked the hair off the floor, and placed it in the small sample bag.

When I first heard of this incident from Bryan Dickeson of UFO Research (NSW), I was naturally suspicious. I wondered if this was some sort of crude hoax. He didn't think so. It was clear to me that the story did not have the strengths of Peter Khoury's case, especially as the chain of evidence—an important factor in forensically establishing context and relationship—had been broken. Mike said he dropped the hair sample, then went back and found it again on the bathroom floor and placed it in the bag. How certain could we be that it was the same hair sample? Mike seemed fairly confident. I wasn't.

When I visited the family, I took control samples from Mike, his partner, his daughter, her daughter, and his son-in-law. On visual inspection, which I then verified under a high-powered microscope, his daughter's hair sample looked similar to the "alien hair." The rest were obviously different. I also determined from Mike that he had read about Peter Khoury's experience a year or so earlier. At the time, he said, he thought it seemed far-fetched, and claimed he didn't even read the whole article. Despite these obvious difficulties of "contamination" by prior information, Mike Wood seemed to be credible. He gave the impression of someone who was struggling to come to grips with what he perceived was a real UFO and alien contact and abduction situation.

Ultimately we did the preliminary step of undertaking a mitochondrial DNA PCR profiling comparison against the "human consensus" on both the "alien hair" and Mike's daughter's sample. While different, both were within the normal "human consensus" DNA range. This "alien hair" sample did not display the unusual variations from normal human consensus in the region of DNA hypervariability seen in the Khoury sample. It had been an informative and worthwhile exercise.[70] We continue to monitor Mike Wood's ongoing experience.

CHAPTER NINE

An Early Abduction Odyssey

IN THE COURSE OF MY RESEARCH INTO THE UFO PHENOMENON, I visited Oliver Harry Turner, a retired Australian government scientist, at his home in the summer of 2001. A nuclear physicist, Turner was the chief health physics officer for the controversial Maralinga atomic bomb trials during the 1950s and early 1960s, but it was his secret and strange odyssey through the UFO mystery that had brought me to his home. He had written a classified "scientific appreciation" report on UFOs for the Australian Department of Defense Directorate of Air Force Intelligence in 1954 that concluded, rather provocatively, "The evidence presented by RAAF [Royal Australian Air Force] tend to support the . . . conclusion . . . that certain strange aircraft have been observed to behave in a manner suggestive of extra-terrestrial origin." During the late 1960s and early 1970s Turner had fought a secret battle to encourage a more scientific and comprehensive examination of the UFO mystery within the clandestine world of intelligence and the military. By then he was a scientific analyst with the Directorate of Scientific and Technical Intelligence of the Joint Intelligence Bureau.[71]

As the focus of my visit was fleshing out Harry Turner's story, we had been discussing his dealings with the military and intelligence on the subject. I was not anticipating the bizarre tilt into the "twilight zone" that was to come. During our discussions he handed me what turned out to be a particularly remarkable document—ten pages of his handwritten notes—devoted to his investigation during 1971 and 1972 of a Canberra woman named Mrs. Klein. I was immediately interested as I knew of a Mrs. Klein, also from Canberra, and I wondered if it was the same woman, as Turner's report told of Klein's extraordinary contacts and abduction milieu. These details revealed an alien odyssey far stranger than the facts I knew about my Mrs. Klein.

The Mrs. Klein I knew was Vicki Klein. I had spoken with her once back in 1991. She had been living with her family in the small town of Dalton, which lies northwest of Canberra, the capital of Australia. At the time, I was examining the possible connection between earthquakes and unusual aerial lights, and this is what drew me to contact Vicki Klein. Dalton has an unusual claim to fame. It is one of the most earthquake-prone areas in eastern Australia. I also knew that in the late sixties Vicki Klein had been involved with a Canberra UFO group. It was that fortunate juxtaposition—a person interested in UFOs now living in Dalton, a locality steeped in earthquake activity—that led me to talk to her. She impressed me as an intelligent and articulate woman, willing to help a researcher examining the "earthquake light" hypothesis for UFOs. She mentioned nothing to me about her own very strange experiences.

That's why Harry Turner's notes on Mrs. Klein's experiences came as such a shock to me. But my shock quickly turned to fascination, as I realized that Harry Turner's 1971–72 notes revealed that Mrs. Klein's bizarre experiences actually prefigured several extraordinary elements of the UFO abduction phenomenon that would not become public for more than a decade and a half, when researcher Budd Hopkins first described them in his 1987 best-selling book *Intruders*. Her story was unpolluted by our modern culture's fixation with the prototypical alien abduction experience. As such, it represents a critical touchstone for investigators of this phenomenon.

Turner's notes revealed that Vicki Klein's first exposure to the UFO mystery apparently occurred in 1960. At the time she was on a CSIRO[72] kangaroo field study with Harry Frith in the far northwestern corner of New South Wales, near Wilcannia, when the two of them observed a red light approaching rapidly. It stopped, changed to blue and then white before shooting up to a high altitude, then falling down again with an odd falling-leaf motion. The light stayed in the area for two hours, during which it repeated these movements about three times. Later in October of the same year a UFO sighting in the southern offshore state of Tasmania by an Anglican minister, Lionel Browning, and his wife, from the rectory at Cressy, gained national publicity.[73] This led Vicki to think that what she and Frith had observed was probably a flying saucer. Frith warned her not to talk about it, perhaps thinking that such an admission might jeopardize his ultimately successful attempts to become chief of the CSIRO's wildlife division. The rest of Vicki's life would prove to be an extraordinary dance with strange phenomena—an ongoing embrace with the mysterious.

Vicki married in late 1960, and three years later she and her husband, Gerry, moved to the then new Canberra suburb of Downer. She worked at a day job while her husband worked the night shift with the Commonwealth police. One night, when she was a few months pregnant with her first son (David), Vicki was sitting alone with her dog—a German shepherd—when suddenly the dog began bristling and snarling at something invisible to her in the corner of the room, preventing Mrs. Klein from going there. On another day the cat suddenly reacted violently in the kitchen. It went through the back door screen and never came back inside the house. Vicki was getting the impression that somebody was watching, but she never saw anyone.

From Harry Turner's notes: "One night by the fire she felt this happen and instead of turning her head as before, she looked out of the corner of her eye to see a man standing in the doorway with his head nearly touching the top (approximately 6'6"). He was dressed in a close-fitting suit of silver mesh." He seemed to radiate a feeling of well-being. After David's birth in 1964, Vicki would often find that his toys had somehow been moved from a lowboy into his cot. Vicki would often see David smile and reach out his arms toward an empty space.

Turner's notes then reveal a dramatic turn: "One day while looking in a mirror Mrs. Klein saw David smile, reach out and then was lifted up from the bed into mid-air." The presence— "the Man"—became known as Claude. When David was about two, "he claimed to have been taken up into the clouds by 'the

Man.' " David had apparently made a sketch of the man's "house," and it appeared to be a rough drawing of a "flying saucer," according to Vicki. Claude's presence in the house persisted for about eighteen months, before disappearing at the time of a "UFO sighting" over Canberra airport.[74] But "Claude" or "the Man" would continue his strange intermittent presence in the Klein household.

Vicki's second son, Adam, was born in 1966. Sometime prior to this Mrs. Klein had her second UFO sighting. She was at the Royal Show in Melbourne, leaning back in the seat of a car, when she saw three objects circling in tight formation. They disappeared when two planes approached their position.

Vicki Klein's third UFO sighting involved an approaching "three-pronged object," when one of the "prongs" seemed to drop off. The remaining two went "into a wobble, approached, and looked like a gray boomerang with a number of black spots down the spine." The object rose three times, on each occasion giving the appearance of "wobbling violently until spinning and then falling." There were two witnesses of Greek background who regarded the aerial apparition as "the work of the devil."

The fourth sighting by Vicki Klein is particularly interesting as it involved setting off a "UFO detector." Harry Turner's notes of his interviews with Mrs. Klein do not record the date, but list the following information: "Cylinder 40°, above Mt. Majura, going [in a rectangular movement, according to Turner's sketch] but drifting slowly northwards. Neighbor as witness." This brief description was supplemented in some detail via correspondence Harry Turner had with Dr. Michael Duggin, a fellow Australian scientist who also had a deep interest in the UFO phenomenon. Harry Turner and Mike Duggin were part of the secret "invisible college" of scientists conducting research in the UFO mystery.

In correspondence with Mike Duggin, Harry detailed Vicki Klein's UFO "detector" sighting. It occurred during a rash of sightings in the Canberra area during June and July 1971. This activity had prompted Turner to contact the local civilian UFO society in Canberra, formed in 1967. Vicki Klein had been the group's secretary since its inception. Turner told me that while Vicki told the group of her sightings, she was careful not to mention the unfolding "alien" dynamic happening in her own house. To her the UFOs were up there and obviously part of the UFO scene; the other events in her home were an entirely different matter. In the wake of Turner's contact with the group, she privately made contact with him and, during a series of meetings at her home between July 1971 and February 1972, Vicki Klein described her unfolding alien contact calling.

At home at 1:10 p.m. on July 7, 1971, Vicki saw a light gray cigar-shaped object above Mt. Majura. Ten minutes earlier, a magnetic detector designed to pick up changes in magnetic fields, and set up inside her house as a crude "UFO detector," started to "buzz" and could not be switched off by resetting. The letter continues, "Armed with 12 power binoculars and a next-door neighbor, she scanned the skies for a possible cause for the buzzing, finally locating an object at elevation 40° and bearing 70°, apparently directly above a civil radar station on top of Mt. Majura (two miles from observer and four miles NE of Canberra city). The object was slightly fatter at one end and was tracing out a square pattern in a vertical plane, the object remaining horizontal during the maneuver. At the corners of the square pattern there appeared to be some fading with occasional flashing. After five minutes the object disappeared and the detector stopped buzzing. There was no echo reported on the radar, but this would not be expected if the object actually was directly above the radar."

Two days later, Vicki Klein spotted another UFO from a sports arena in the afternoon. Schoolboys playing on the field alerted her to its presence. It seemed to be a dark gray object without wings, tail, or contrail, moving slowly over Mt. Majura. The object was about half-moon size, with a silver shimmer at the rear.

Harry Turner made the following notes after a July 1971 conversation with Vicki: "For two years she had been instructed in the astral by a being dressed in a tight fitting uniform and with a head more like a dog (pointed ears etc.).[75] For some months, local individuals were brought to her house for contact, and then astral instruction by her. They (UFO people) are too busy to do everything themselves and need earth teacher." This was the beginning of a contact group that formed around her.

Vicki elaborated on her understanding of the alien contact. "Children even in embryo are being subconsciously trained for the latter days. Some children are being 'fathered' by space-beings (by non-physical means). [Vicki] believes that Claude was responsible for her second child Adam who was generated at about the time Claude returned from an absence and while she was still on the pill. Information is being stored in their minds for use at an appropriate time. She believes this has also happened to her."

The "space beings" told Vicki that "the latter days" were a coming time of chaos. She told Turner: "The signal for the beginning of the end will be the disappearance of a small island followed by a larger one—she suspects the New Guinea area. The chaos itself will last three days getting progressively worse. The world has to be cleansed, and a new development will take place in lands newly arisen." This is a very familiar mantra in UFO contact experiences—contact and prophecy—usually rooted in something apocalyptic and generally always unrealized. It is fascinating to see how these flawed belief systems are so easily and uncritically embraced.[76]

The impending "chaos" to be played out on Canberra—including the convoluted "signs" heralding it, the preparation for it, and its exact nature—became a major preoccupation for Vicki Klein and Mrs. Z., one of her neighbors. Turner's notes reveal: "Both women claim they have been told that when the chaos comes they are to go to a particular location south of Canberra with their children and await events. They already have some food prepared to take with them. The first day of the chaos will be severe earth tremors, weird colors in the sky, green discs flying over and red balls fired at the ground, which will start fires. Later (second or third day) will come the major earthquakes. . . . Their job is to look after the children. Mrs. Z. has an adopted child, who she considers is really one of the blonde race's children. . . . David is Mrs. Klein's eldest and is not one of 'theirs,' whereas Adam the second eldest is really 'Claude's.' They have been told that they will have more children even though Mrs. Z. has had a hysterectomy and Mrs. K has had her tubes tied.[77] Their purpose is to bring up these children so that they can go to the new land when it arrives. . . . The coming chaos will be used by these people to bring about a new earth race. One difficulty appears to be that there is an evil group opposing this plan and anxious to take over the earth for their own purposes. This other group is attacking and converting people who are actively helping the first group. This struggle will be even greater during and after the chaos. Apparently, both groups can be rather forceful in trying to keep their earth protégés in line. The 'good' side uses the excuse that they only exert influence on their own race descendants. . . .

"It appears that mainly women are saved from the chaos, presumably to look after the children. No men have been seen in the valley where the children are later taken. Mrs. Klein considers that she herself is a 'space-child' and is 'one of them.' She has been told that if the natural chaos is insufficient for the task, then additional chaos will be created. There is a determination to do the job. There will be a great 'battle' with the evil forces who will try to take over. In fact, they believe the evil ones will hunt them down after the chaos. Not only will there be eruptions among the hills of Canberra, but there will be much water in some way. The space people will be so busy during this time that they cannot give assurance of help. They may be able to provide a 'lift' to the hiding place, but will not be able to stay afterwards. There is no interest in helping the men in the families."

If this were the extent of Vicki Klein's story, it would be of doubtful interest. It's what follows in Harry Turner's notes that really struck me—how the story strikingly resembles the alien abduction accounts that would come to dominate the UFO controversy fifteen years later. Turner wasn't sure what to make of this information in the early seventies. At the time he was simply trying to make an accurate report of Vicki Klein's claims.

Of the two Canberra women, Vicki Klein and her neighbor Mrs. Z., Turner writes: "They both claim to have had ova taken from them (they say the operation is quite painful) and that children are being created and bred in some other place for the time when they will be returned to earth. They both claim that on the same night they were both taken astrally to a place, a huge room whose walls were covered with small cells, Mrs. Z. saw embryos being incubated. In the centre of the room, Mrs. Klein saw a central glassed enclosure containing developing babies that had been taken from the incubators. Many of them did not look well. When the 'nurse' was not looking, Mrs. Klein, disobeying instructions, took one from its humidity crib and nursed it. The attendant was at first very annoyed, but later realized that the baby was more contented and that 'motherly love' was the missing ingredient. In fact this lack of emotion and feeling of the advanced race is one of its downfalls, because there is no ambition to perpetuate the race. One of its present aims is to combine the advanced intellectual and psychic abilities of its race with some of the emotions of the human race. Hence these cross-fertilized embryos."

With regard to the use of human ova in the aliens' experiments, Turner notes that "Mrs. Klein said that after she had complained of the pain caused by the 'operation,' during an astral projection she was invited by a friend to have intercourse, but she refused, whereupon the friend changed into her astral guide, but she still refused. The story was given to me in case it could help in understanding their experimental process. She thought the offer was made as an alternative method of producing their babies without the pain of the operation. Alternatively, it may have been a test of her 'moral fiber.' Since then, she has had many ova removed. Once again, she insists that these people are not interested in earth people, but only in reclaiming 'their own.' She does not know what is meant by 'their own.'

"They communicate with Mrs. Klein by producing a high-pitched buzzing noise that rises and falls in intensity at about once a second, followed by a voice which is external to the ear.

She communicates by repeating speech or thoughts—they apparently are not monitoring her continuously—until an answer is received (usually a day or two). In the presence of the two types, only the 'dog-faced' one converses, but even then his mouth does not move, although the voice seems to come from him. He has black markings over his eyes and his black pupils are like tunnels. His hair is dark; there is an overall bluish tinge. The blonde one only communicates reassurance."

Mrs. Z. claimed that during the evening of September 7, 1971, she had gone to the rear of the house because her dog was acting up. Turner's notes state: "As the dog retreated indoors, she noticed two figures in Mrs. Klein's backyard near a tree. Mrs. Z. thought they must be Mr. and Mrs. Klein but was puzzled as to why they should be standing in that position and be so still. So she went towards them and shone her torch [flashlight] on them. They turned out to be a tall metal-meshed clad figure (like 'Claude') with his 'dog-faced' companion. They were smiling at her as though a joke was being played on her and then gradually faded away. No signs remained. Mrs. Klein's dog was not disturbed.

"These two disparate types claim they are 'one people.' The tall blonde one is similar to the Nordic race, but the 'dog-faced' one has pointed ears, eyes with large pupils, and no iris (or is it no white —Turner), longish flattened nose, thin mouth, dressed in blue-grey tunic with tight banded collar and oval (?) shoes. The dog-faced type can appear to look more human-like if they wish to (swarthy, 5'–5'6")."

In this account we have several features that are identical to the ones heralded as a breakthrough for the understanding of the alien abduction agenda by Budd Hopkins in 1987: the humanoid beings, one of them apparently of the "Nordic type," involved in a genetic program that utilizes painful ova removal procedures, taking place in a strange room that houses bred babies (an alien incubatorium), and babies that seem weak and respond to bonding with humans. Even Mrs. Klein's being encouraged to have sex, only to find that the encouragement is actually coming from one of the entities, is a feature that would be described by another researcher, David Jacobs, in *Secret Life* twenty years later as an "envisioning" procedure. The resemblances are astonishing.

For Vicki, the "contacts," strange "visions," and unusual phenomena continued to escalate in the latter part of 1971. But the encounters were not always clear-cut. On the night of September 12, "the house was full of friendly observers, but no communication was made. Often there is a knock at the front door, but no one is there. She considers that it is their polite way of gaining entry. Sometimes shadows are seen, maybe across a window or perhaps a disturbance of light, that looks like a shadow, but isn't, crosses a room." The visions became darker, and she was often taken on "projection"-type astral trips, where scenes seemed to be played out in front of her as if on a screen. Many of these displays seemed to be of events to come—a kind of reinforcement of prophecy. Vicki did not volunteer what she was going through nor that fires were connected to the coming "chaos." Things continued into early 1972 with the feeling that events might come to a head by February 22. Both of Vicki's sons—David and Adam—were having dreams of fire, and talked of being places with "the man." During storms in February Vicki Klein sensed many presences, and saw one of "them." "She is to clear away items in the backyard to provide a landing space. Adam awoke one morning and said that he thought his mother had already left with 'the man.'"

Harry Turner's notes end at that point. He lost contact
with Vicki Klein and her neighbor. Needless to say, earthquakes, fire, and water did not decimate Canberra in 1972.[78]
The spiral of visions and contacts that Vicki Klein had been
sucked up into seemed to come to nothing.

When Harry Turner passed on this material to me in July
2001, I immediately set about trying to confirm its contents.
There were many problems. The worst was that Vicki Klein
had passed away in 1994. In a further tragic turn, I discovered
that her eldest son, David, had died in June 1978 at the age of
13, having fallen out of a tree. In February 2002 I visited Canberra, its suburb of Downer, and then went out to Dalton,
where the family had lived since 1972. Some locals gave me
some valuable background. One of them, Wayne Medway of
the local garage, grew up in the same time frame as Adam
Klein. He always thought Vicki was okay; she was certainly
not "off her perch." He was aware of Vicki and Adam's UFO
interests and even recollected there was something about a
UFO landing in the paddock near the Klein house.

I then learned from Wayne that Vicki's father, Ron Summers, still lived there. Ron graciously agreed to an interview.
He confirmed the basic story that Vicki had revealed to Harry
Turner. While Ron didn't quite know what to think about
them, he basically accepted his daughter's strange experiences. He even witnessed UFOs with her one night, so he
didn't discount the subject. Despite all this, he felt that Vicki
had her feet planted in the real world as well.

In April 2002 I located Vicki's sister, Gillian. She remembered the talk about "Claude," but she was never sure about
"the alien thing." As for Adam being an "interloper," understandably she was very skeptical about that. For Gillian,
Adam looked and acted like a member of the Klein-Summers

family tree. She recollected that the three children, David, Adam, and their sister were all firm believers in UFOs, at least when they were young and living in Canberra and Dalton. She remembered that Vicki had a very strong affinity with animals. Gillian had been reading a series of novels anchored in Native American spirituality that featured "dreamers," or shamans that could foresee things. They reminded her a lot of Vicki—a shamanic "dreamer." In truth Gillian said that she and some members of her family seemed to have been in contact with Vicki since her death on a number of occasions, often sensing her around them. In life Vicki was full of contradictions, but she managed to knit it all together. Basically the whole family just accepted Vicki. None of them seemed to think anything was wrong with her. According to Gillian, even Gerry, Vicki's husband, who was very pragmatic and skeptical of such things, would just shake his head and say, "That's Vicki."

I spoke with Gerry. Although guarded, he said, "I wasn't keyed into it, Vicki would say. I was not a believer." Realizing that these things didn't interest him, Vicki did not share everything with him. She saw him as a psychic block of wood, not open to the wider, stranger aspects of reality. He claimed that having horses and a few acres of their own was their main motivation for moving to Dalton, not concern about a catastrophe in Canberra.

I finally spoke with Adam Klein, Vicki's second son, then 36. He relaxed and warmed to the conversation when he realized I wasn't a wild-eyed believer who saw him as an alien child, and that my interest was more sociologically and scientifically motivated. From his recollection and perspective he said that his mother had "vision," and that she had a deep spiritual side, which encouraged psychic development in her children as part and parcel of childrearing. He said that the reason for the family move from Canberra to Dalton was strongly tied up in Vicki's fixation or fear of a looming catastrophe (such as "the earthquake") smiting Canberra. Adam recollects that the family went along with it, but never believed any of it. He felt sorry for his father, as he had a good job in Canberra and was forced to go out to "the middle of nowhere" and work on the local Gunning Council in a job he never really liked. Adam said his brother David was a lot closer to his mother. I was surprised when Adam confirmed something that Vicki's sister told me: that Vicki's "spirit" had gone to one of her sister's daughters. Family lore suggested there was now some sort of psychic link between generations. I thanked Adam for his time. In Australian parlance, he was a nice bloke. He even volunteered a DNA sample, but probably hopes I'll never come asking for it.

Before leaving the Vicki Klein story, we should look at the Dalton connection. While living there, Vicki kept the Sydney-based UFO Investigation Centre apprised of some of the stranger goings-on there. Some surely owe their origins to the earthquake proneness of the area.[79] In one undated Dalton dispatch, probably from the late seventies, Vicki wrote, "Something strange happened two nights ago. I woke up at 2.30 a.m. with an awful shriek because I had walked in my sleep into the wardrobe door, with some force too. I have a large bruise on my arm. I felt as if I was drugged and I really fought to clear my mind, it was something outside.

"There was a terrible racket going on, dogs everywhere were howling, and the horses were whinnying and racing around the paddock, birds screaming and cows moving and also racing around.

"I went outside. There was no moon, but bright starlight.

No wind. I couldn't see anything, but there was a 'singing' which seemed to fill the air and go right through me. That's the only way I can think to describe the noise.

"After twenty minutes it suddenly stopped and all the animals quieted down. I went back to bed and slept till morning. I have since asked eleven people, without explaining why, about that night. Five had walked in their sleep and all said they had terrible dreams, but couldn't remember them except they were very frightened."[80]

It's worth noting that Vicki took up a serious interest in earthquakes in the Dalton area.[81] I first realized this in 1988 when I saw a segment on the popular science show *Quantum* on the ABC (Australian Broadcasting Corporation) TV network. It was a story about Dalton and how scientists were working with locals to monitor seismic activity. The key local was Vicki Klein. One of the scientists was Dr. Marion Leiba, of the Australian Geological Survey Organisation. She was one of the first people I called after I received the Turner notes on Vicki Klein. I wanted to find out as much as I could about Vicki. I soon found that Marion Leiba had a very high opinion of Vicki, and they had become firm friends. I met with Marion at her Canberra home and told her the details of Vicki's strange story, which she had not revealed to Marion. Despite the bizarre nature of the story, Marion insisted that the Vicki she knew well was entirely sane and a good friend. Marion was aware that Vicki had some unusual and perhaps eccentric interests, but none of this diminished Vicki's stature in her eyes. When Vicki died of a brain hemorrhage on April 21, 1994, Marion wrote an obituary about her friend for an in-house geophysical survey/engineers newsletter.

Vicki Klein's story prefigured one other provocative aspect of the UFO abduction phenomenon, and that is the notion of "star children."[82] Some believe that "special children" with enhanced abilities are the product of human-alien unions or other exotic means and that they represent a transformational step in mankind's artificially accelerated evolutionary path. Stories like Vicki Klein's might suggest such a notion, but they are far from being the kind of proof that the "star children" saga requires. An uncritical fusion of New Age perspectives has crept into the unraveling of such claims. While some advocates of "star children" may mean well, and are perhaps only seeking to assist the "alienated" achieve their full potential, it is a veritable mine field that requires critical and objective assessments. Such precautions are absolutely paramount in dealing with children, and yet the opposite seems to be happening. It is one thing for parents of "special children" to hold such belief systems, but it is quite another for them to foist such beliefs onto their children. Those parents—and researchers—who do, run the risk of being accused of exploiting or endangering children. We need frequent reality checks as we work toward answers to these strange and puzzling questions.

CHAPTER TEN

Shaman Blues

TO THE NAMES PETER KHOURY AND ANTONIO VILLAS-BOAS, add the name Credo Mutwa. Credo Mutwa is a *sangoma* (a shaman or healer) and high *sanusi* (clairvoyant and lore-master) of considerable repute. Born in 1921 in the Natal area of South Africa, his full name is Vusamazulu Credo Mutwa, which is the product of a mix of cultures, essentially meaning "Great Awakener (of the Zulus), I Believe (in) the Little Bushman." He is also a noted painter and writer whose published works include *Indaba, My Children: African Tribal History, Legends, Customs, and Religious Beliefs* (1964), *Africa Is My Witness* (1966), *My People* (1971), *Let Not My Country Die* (1986), and *Isilwane: The Animal: Tales and Fables of Africa* (1996). Although its accuracy is often disputed, *Indaba* is epic in its scope and often referred to as a "classic."

As a shaman and keeper of native stories and traditions,

Credo is held in high regard by many within the community of academics and researchers who examine myths, native healers, and shamans. Dr. Stephen Larsen[83] edited Credo Mutwa's *Song of the Stars: The Lore of a Zulu Shaman* (1996) and Dr. Bradford Keeney[84] edited *Vusamazulu Credo Mutwa: Zulu High Sanusi* as part of a series of books called *Profiles of Healing*, which are dedicated to "helping the world's most revered healers, shamans, medicine people, and leaders of complementary medicine tell their stories." Credo is recognized as "the most famous African traditional healer of [the twentieth] century," "a national treasure of South Africa," and an "elder spokesman for the African healing arts."[85] With this sort of background of respect and regard, it seems surprising that Credo Mutwa would embrace the alien controversy. But he does.

In fact, he claims that he was the victim of a horrific alien abduction and seduction back in either 1958 or 1959. Credo's experiences have been detailed in several books and at least one video, but for this account I quote with permission from an interview he gave to Rick Martin, editor of the alternative newspaper *Spectrum*, in 1999.[86] A warning in advance: Parts of this story are extremely unpleasant and other parts are quite explicit. Interspersed with the extensive quotations from Credo (in italics) are my own comments.

Credo Mutwa begins by telling the story of how he got "taken."

Sir, it was an ordinary day, like any other day. It was a beautiful day in the eastern mountains of Zimbabwe, which are called Inyangani. *These are mountains to the East of Zimbabwe.*

Now, I had been instructed by my teacher, a woman called Mrs. Moyo [Mrs. Zamoya], was Ndebele, from Zimbabwe, once known as Rhodesia, to go and find a special herb, which we were going to use in the healing of a certain initiate who was badly ill.

I was looking for this herb, and I was not thinking about anything, and I had no belief whatsoever in these creatures. I had never encountered them before, and although we African people believe in many things, I was mighty skeptical, even about certain entities that we believed in at that time, because I had never encountered anything like that before.

Presumably here Credo Mutwa is referring to alien abductions as distinct from UFO encounters, for in *The Song of the Stars* he describes some UFO experiences that took place in Botswana between 1951 and 1958. Credo claims that in 1951 he witnessed a UFO, two entities that ran into the object, its subsequent departure, and the hazardous physical ground trace it left behind. He reports that in 1958, he was called in as a *sangoma* to look into a bizarre event that involved his encounter with a monstrous cylindrically shaped creature known as the *Muhondoruka*.[87]

And all of a sudden, sir, I noticed that the temperature around me had dropped, although it was a very hot African day. I suddenly noticed that it was now cold and there was what appeared to be a bright blue mist swirling all around me, getting between me and the eastern landscape. I remember wondering, stupidly, what this thing meant, because I had just begun to dig one of the herbs I had found.

Suddenly, I found myself in a very strange place, a place that looked like a tunnel lined with metal. I had worked in mines before, and where I found myself appeared to be a mine tunnel, which was lined with silver-greyish metal.

I was lying on what appeared to be a very heavy and very large working bench or a working table, sir. But yet, I was not chained to the table. I was just lying there and my trousers were missing, and so were the heavy boots that I always wore when I was out in the bush. And all of a sudden, in this strange, tunnel-like room, I saw what appeared to be dull, heady-looking, grey, dull-like creatures, which were moving toward me.

There were lights in this place, but not lights as we know them. They seemed to be patches of glowing stuff. And there was something above the far entrance, which looked like writing, that writing against the silver-grey surface, and these creatures were coming at me but I was hypnotized, just as if the witchcraft had been put upon my head.

But I watched the creatures as they were coming towards me. I didn't know what they were. I was frightened, but I couldn't move my arms or my legs. I just lay there like a goat on a sacrificial altar. And when the creatures came towards me, I felt fear inside me. They were short creatures, about the size of African Pigmy. They have very large heads, very thin arms, and very thin legs.

I noticed, sir, because I am an artist, a painter, that these creatures were built all wrong from an artist's point of view. Their limbs were too long for their body, and their necks were very thin, and their heads were almost as large as full-grown watermelons. They had strange eyes, which looked like goggles of some kind. They had no noses, as we have, only small holes on either side of the raised area between their eyes. Their mouth had no lips, only thin cuts as if made by a razor.

And while I was looking at these creatures, sir, in amazed fascination, I felt something close to my head, about my head. And when I looked up, there was another creature, a slightly bigger one than the other, and it was standing above my head and was looking down at me. . . .

I looked into the thing's eyes and I noticed that the creature wanted me to keep looking into his eyes. I looked and saw that, through these covers over their eyes, I could see the creature's real eyes behind this black, goggle-like cover. Its eyes were round, with straight pupils, like those of a cat. And the thing was not moving its head. It was breathing; I could see that. I could see little nostrils moving, closing and shutting. . . .

The creature smelled like nobody's business. It had a strange smell, a throat-tightening, chemical smell, which smelled like rotten eggs, and also like hot copper [sulfur], a very strong smell.

*And the creature saw me looking at it, and it looked down at me
and, all of a sudden, I felt a terrible, awful pain on my left thigh, as if a
sword had been driven into my left thigh. I screamed in pain, horrible,
calling out for my mother, and the creature placed its hand over my
mouth. You know, sir, it was like—if you want to know how that felt,
please sir, take the leg of a chicken, a live chicken, and place it against
your lips. That was how the creature's hand felt upon my mouth.*

*It had thin, long fingers, which had more joints than my human
fingers have. And the thumb was in the wrong place. Each one of the
fingers ended in a black claw, almost like certain African birds. The
thing was telling me to be quiet. And how long the pain went on, sir, I
don't know. I screamed and I screamed and I screamed, again.*

*And then, all of a sudden, something was pulled out of my flesh,
and I looked down and saw my thigh covered with blood, and I saw
that ōne of the creatures—there were four of them, other than the one
standing over my head—they wore tight fitting overalls, which were
silvery-grey in color, and their flesh resembled the flesh of certain types
of fish that we find in the sea off South Africa. And the creature standing above my head appeared to be a female. It was somehow different
than the others. It was taller, bigger. Although it didn't have breasts like
a woman, it appeared to be feminine. And the others appeared to be
afraid of it. I don't know how I can describe this.*

The sense that a being is feminine, despite the absence of
obvious gender differences, has been reported before. This element features prominently in Whitley Strieber's experience
described in *Communion*. It also occurs in the Australian case
that Strieber felt was the best description of the "grays" he
had ever received.

*And then, while this terrible thing was going on, another of the
creatures came up to me—it walked sideways, in a slightly jerking
way, as if it was drunk—it walked up along the table, to my right side,
and it stood next to the one standing above my head. And before I*

knew what was happening, this creature stuck something that was like a small, silver, ball-point pen with a cable at one end, it pushed this thing, coldly, into my right nostril.

Sir, the pain was out of this world. Blood splattered all over. I choked and tried to scream, but the blood got into my throat. It was a nightmare. Then, it pulled the thing out and I tried to fight and sit up.

The pain was terrible, but the other thing above my head placed its hand upon my forehead and kept me down with very little force. I was choking and trying to spit out the blood, and then I managed to turn my head to the right to spit out the blood, which I did, and then what the creatures did to me, sir, I don't know.

All I do know is that the pain went away, and in place of the pain, strange visions flooded my head, visions of cities, some of which I recognized from my travels—but, cities which were half-destroyed, the buildings having their tops blown away, with windows like empty eye-sockets in a human skull. I saw these visions again and again. All the buildings that I saw were half-drowned in a reddish, muddish water. It was as if there had been a flood and the buildings were sticking up out of this great flood, partly destroyed by a disaster of some kind, and it was a terrible sight.

The "alien" display of scenes of destruction in abductee accounts is common. Vicki Klein reported them in her "alien prophecy" tableaux.

And then, before I knew it, one of the creatures, the one standing next to my feet, drove something into my organ of manhood, but here there was no pain, just a violent irritation, as if I was making love to something or someone.

And then, when the creature withdrew the thing, which was like a small, black tube, which it had forced into my organ of manhood, I did something which produced a strange result, and I did not do it intentionally. I think it was—my bladder opened, and I urinated straight into the chest of the creature, which had pulled the thing out of my organ.

And if I had shot the creature, it would never have reacted as it did. It jerked away and nearly fell, and then it recovered and staggered away like a drunken insect, and left the room. I don't know whether my urine did it; I don't know . . .

Then, after a while, the other creatures went away, leaving me with a dull pain in my nostril, with blood on my thigh, and the table wet with urine. And the thing standing above my head had not moved. It just stood there with its right hand touching its left shoulder, in a strangely beautiful and feminine way. It stood there looking at me. There was no expression in its face. I never saw any of the creatures talk or make any sound of any kind. All I do know is that they appeared to be mute.

And then, out of somewhere there arrived two other creatures, one of which was made entirely out of metal. Even in my worst nightmares, I still see this creature. It was tall. It was big. And the area in which we were was too small for it. It walked with a slight stoop, moving forward, and it was definitely not a living thing. It was a metal creature, a robot of some kind. And it came and it stood near my feet, its whole body clumsily bent, looking down at me. There was no mouth. There was no nose. There were just two bright eyes, which seemed to change color, and seemed to move somehow, like the crackling of an electrical device.

Accounts of metallic beings occur often in the literature. Again, Strieber reported such creatures in *Communion*.

And then, behind this huge, bent creature, came a creature which surprised me. It was very, very, very, very swollen, sir, in appearance. It had pink skin. It had a blondish, very human body. It had very bright, blue, slanting eyes. It had hair which looked like nylon fiber of some kind. It had high cheek-bones and an almost human mouth, with full lips and a small, pointed chin. The creature, sir, was definitely a female but like an artist and a painter, which I am, and also a sculptor, I noticed that the creature was totally out of proportion. It was wrong.

First, its breasts were thin and pointed, and set too high upon its chest, not where a normal woman's breasts would be. Its body was powerful, almost fat, but its legs were too short and its arms were too short in proportion to the rest of its body. And it came towards me, looked down at me, and before I knew what it was doing, somehow it mated with me. It was a horrible experience, sir, even worse than what had been done to me before. But even now, the trauma of that day had affected my life even now, exactly forty years later.

Why did that strange creature, which was naked, with red pubic hair, which climbed over me on that working table, why did it have an organ, which though slightly different from that of a normal woman, was still a recognizable female organ?

In an interview with controversial researcher David Icke,[88] Credo Mutwa elaborated on the description of this strange female. He said her breasts were set high. Her eyes were pale blue and her head hair a golden color. But her pubic and armpit hair was fiery red in color. The contrasting hair color is strikingly reminiscent of the female being encountered by Antonio Villas-Boas only a year or two earlier. Credo Mutwa also noted that the female creature had a tail-like appendage. Was this a genuinely recollected detail or an embellishment inspired by an awareness of Villas-Boas's story? When I spoke to Credo Mutwa in 2002 I did not get the impression that these elements were inspired by the 1957 Villas-Boas story. The often poignant nature of our discussion left me with the impression that this was not about recent elaborations and fabrication, and perhaps was more to do with a real story told forthrightly but influenced by the crosscurrents of his complex life journey.

The creature's organ was in the wrong place. It was slightly more in the front, where that of normal women is between the legs. But it was recognizable, and it looked like a female organ. It had hair like a woman's organ.

And after that, when the creatures had gone, leaving only the one creature which had been standing about my head, the creature standing about my head shook me by the hair, it gripped me by the head and forced me to stand off the table and to get off the table. I did that, and such was the state that I was in, that I fell onto my knees and hands, onto the floor.

And I noticed that that floor was strange. It had moving patterns in it, which kept on changing and shifting—purple, red, and greenish patterns, on a metal-grey background. And the creature pulled me by the hair, again, forcing me to stand up, and it pushed me roughly and made me follow it.

Sir, it would take too long for me to describe what I saw in that strange place, as the creature pushed me, roughly, from room to room. Even now my mind can't grasp what it was that I saw. Amongst many things that I saw were huge cylindrical objects, made of what appeared to be glass of some kind. And in these objects, cylinders, which reached from the roof to the floor of the place we were going through, was what appeared to be a sort of a greyish-pink liquid. And in this liquid I saw small editions of the alien creatures floating round and round, like disgusting little frogs, inside this liquid.

I couldn't understand what it was that I was being shown. But then, in the last room I was led through, I saw people, and other strange creatures, which, even now, my mind can't make head or sense out of, lying on the table.

And I passed a White man, a real White man, who smelled like a human being, was smelling of sweat, urine, excrement, and fear. This White man was lying on a table like the one I had been lying on, and I looked into his eyes and he looked into mine as I went by.

Abductees in the West also mention the presence of other helpless abductees aboard the "ship."

And then I found myself out in the bush. I found that my trousers were missing. There was a terrible pain in my left thigh. There was a *pain in my penis which was starting to swell, and when I tried to pass water, the pain was excruciating.*

Peter Khoury reported the same problem following his July 1992 experience.

I took off my shirt and I used it as a loin-cloth and I walked through the bush.

I first met a group of young Rhodesian Black people who guided me to my teacher's village. And when I arrived outside that village, I smelled so horribly that every dog in the village came yapping and snarling to tear me to pieces. And it was only my teacher and her other students and the villagers who saved me on that day. My teacher and the villagers were not at all surprised by what I had to tell them. They accepted it, sir. They said to me that what had happened to me had happened to many other people before, and that I was lucky to return alive, because many people have disappeared in that part of the land, never to be seen again—White people, Black people, and so on.

Credo Mutwa's teacher, Mrs. Zamoya, told him that he had been missing for three days.[89] Arizona abductee Travis Walton went missing for five days after a UFO encounter.[90]

In the year following, 1960, I was delivering parcels in the city of Johannesburg. You see, I was working in a curio shop, when a White man shouted at me to stop. I assumed that the White man was a secret policeman who wanted to look into my identity documents. And when I tried to produce the documents, he told me, angrily, that he didn't want to see my stinking documents.

Sir, he asked me this question: "Listen, where the hell have I seen you before? Who are you?"

I said, "I am nobody, sir; I am just a working man."

He said, "Don't bullshit me, man; who the hell are you? Where did I see you before?"

And then I looked at him. I recognized him—his long, straggly, golden-brown hair, his ridiculous mustache and beard. I remembered him—his blue eyes blood-shot and naked-terror, shining upon his eyes, and his skin as pale as that of a goat.

I said, "Meneer," which is the African's way. "Meneer—I saw you in Rhodesia in a certain place underground." And if I had hit that White man with my fist, he wouldn't have reacted the way he did, sir. He turned away and walked with a terrible expression, and he disappeared on the other side of the street . . .

Over many years of looking into this thing, trying to understand it, I can tell you this: that the Mantindane, *and the other kinds of alien beings that our people know about, are sexually compatible with human beings.*

This theme essentially entered the ufological canon in 1987. Budd Hopkins, in his books *Intruders* and *Missing Time,* best expressed the mainstream view of the American UFO community regarding alien abductions. The alien abductors, in his view, are extraterrestrial and are involved in some sort of extraordinary genetic experiment. Human beings are first abducted when children. Their cells are sampled, and as a result, certain individuals are followed closely. After puberty, ova and sperm cells are taken from them during follow-up abductions. Sometimes pregnancies follow that later seemingly mysteriously abort. The goal of this activity is "the merging of human and 'alien' genetic material for the production of a hybrid race." The hybrids are then apparently brought to term in laboratory "nurseries" inside large UFOs. The involuntary contributors are later abducted yet again, and are shown their offspring—tiny hybrid infants or children. In these bizarre "baby presentation" events abductees are asked to pick and hold their "offspring" in a kind of bonding experience.

The Mantindane *are capable of impregnating African women. And I have come across many cases of this during the last 30 years or so. For example, according to our culture, abortion is regarded as worse than murder. And if a tribal woman from a rural area in South Africa is found to be pregnant by some unknown person, and then her pregnancy disappears, [the] relative to that woman, accuses her of having committed abortion, and yet she denies this, of course. And because of the fight that results between her and her relatives, the husband's relatives, then she challenges these people who are accusing her to take her to a* sangoma; *that is a person like myself. The* sangoma *will sometimes examine the woman and, if the* sangoma *finds that the woman had been pregnant, and had somehow had her fetus removed—a thing which, when it is done by the* Mantindane, *results in specific injuries to the woman which anyone with experience can recognize—then, the* sangoma *knows that the woman is telling the truth. Also, the smell which clings to people who have been through the hands of the* Mantindane, *that meticulous man which is unforgettable, always clings to all women who have been impregnated by the* Mantindane, *no matter how much perfume or powder they try to use.*

So, in the last 40 years or so, I have received many women who have actually been impregnated by the Mantindane *and their pregnancies mysteriously terminated, leaving the woman feeling defiled, feeling guilty, and rejected by her family. It becomes my duty to convince the family of the woman's innocence, to try and heal the terrible spiritual and mental—as well as physical—trauma that the woman has undergone, and to otherwise help her and her members of the family, and forget what happened.*

Other indigenous people have reported similar experiences. For example, John Kernott, who had a mail run in the Kimberleys area of northwestern Australia, had heard a lot of strange stories from aboriginals. He told me that tales of sexual contact resembling alien abductions were apparently very common among aboriginal women in the Great Sandy and Central Deserts. Since about 1986, Central Desert aboriginal women have told him of "feather foot" spirits that do not walk on the ground but walk in the air and move through walls and things in order to punish people. They talk about waking up and having sex with spirits, having babies, though no baby ever comes. Often their stomachs swell up and then go down. Their sexual assailants are often described as being like "men in black." It was often hard to see their eyes. "Phantom" or "false" pregnancies, or missing fetuses feature in UFO abduction lore. The concept of "spirit children" is also widespread. Such comparisons are fascinating, but ultimately they are difficult to interpret, given the cultural filtering and distortion that is likely in such a fringe area.

John's father-in-law, an aborigine, told him that as a young man in the 1930s, while camping with a group of aborigines, they saw a green light spinning around in the sky. It landed behind trees. Lots of little men shining with green light came and walked around the aborigines, looking at them, and then walked back to the UFO, which then took off.[91]

I believe that these so-called aliens don't come from far away at all. I believe that they are here with us, and I believe that they need substances from us, just as some of us human beings use certain things from wild animals, such as monkey glands, for certain selfish purposes of our own.

Credo Mutwa was apparently not the only person to disappear for days under unusual circumstances from Mt. Inyangani. A recent deputy minister in the Zimbabwean government was one of three companions "lost" on the mystery mountain. What was truly bizarre was that according to the minister the three men "walked aimlessly in a state of confusion, feeling neither thirst nor hunger, all the while seeing and waving frantically at the elements of the rescue team, who could not see them at all." Following ritual magic offerings to "the tutelary deities of the mountain," the men were finally found. This anecdote conjures up a sense of rifts in our space-time continuum.[92] On another occasion a government official went missing on the mountain. Once again, after local tribal ritual efforts, the man was found fit and well, but unable to remember what had happened to him during the two days he was missing. Others were apparently less fortunate and have apparently not been found.[93] While Credo Mutwa went missing on Mount Inyangani he remembers at least part of his three days in alien oblivion.[94]

Far too many people fall into the temptation of looking upon these "aliens" as supernatural creatures. They are just solid creatures, sir. They are like us; and, furthermore, I'm going to make a statement here which will come as a surprise: the Grey aliens, sir, are edible. . . . Their flesh is protein, just as animal flesh on Earth is, but, anyone who ingests Grey alien flesh comes very, very close to death. I nearly did.

You see, in Lesotho there is a mountain called Laribe; it is called the Crying Stone mountain. On several occasions, in the last 50 years or so, alien craft have crashed against this mountain. And one last incident was reported in the newspapers not so long ago.

An African who believes that these creatures are gods, when they find the corpse of a dead Grey alien, they take it, put it in a bag, and drag it into the bush, where they dismember it and ritually eat it. But some of them die as a result of ingesting that thing.

About a year before I had the experience from the Inyangani Mountains, I had been given, by a friend of mine in Lesotho, flesh from what he called a sky god. I was skeptical. He gave me a small lump of grey, rather dry stuff, which he said was the flesh. And he and I and his wife ritually ate this thing, one night. After we had eaten this thing, sir, on the following day, exactly, our bodies erupted into a rash, which was like nothing I had experienced in my life before.

Though considered bizarre by Western standards, the ingestion of flesh and other materials to gain knowledge is a prominent aspect of many tribal cultures.[95] In addition, the symptoms Credo Mutwa describes after ingesting the alleged alien flesh seem rather like a form of anaphylactic shock. Though this hardly contributes to establishing credibility in Western eyes, it is not "alien" to shamanic and tribal lore. If the stories have any substance within an African tribal setting, the "alien flesh" is valued in ritual. As a chemist and one trying to promote biochemical DNA verification of claimed abduction evidence, I think it would be interesting to get some of this "alien flesh," but I suspect that it would be impossible to verify the chain of evidence, even given the remote likelihood of any of this material surfacing. I suspect that like many Western stories of alleged biological evidence in abduction or alien tales, they will remain apocryphal and unverifiable. But such difficulties do not dissuade me from trying to look into these unlikely avenues of evidence.

Our bodies were so full of the rash and urticaria, it was as if we had small pox. We itched, the itching was horrible, especially under the arm-pits and between the legs, and the buttocks. Our tongues began to swell. We could not breathe. And for a number of days, my friend, his wife and I were totally helpless, secretly attended by initiates who were studying under my friend, who was a shaman.

I came very close to death. There was bleeding from nearly every orifice in our body. We passed blood, much blood when we went to the toilet. We could barely walk, barely breathe. And after about four or five days, the rash subsided, then the peeling of the skin took its place now. Our skins began to peel, in scales like that of a snake shedding its skin.

Sir, it was one of the most terrible experiences I had undergone. In fact, when I began to feel better, I think that my being abducted by the Mantindane *was the direct result of my having ingested flesh from one of these creatures. I had not believed that what my friend was giving me was flesh from a creature. I assumed it was some kind of root or herb or whatever. But, afterwards, I recalled the taste of the thing. It had a coppery taste, and had the same type of smell that I was to encounter in 1959.*

It is intriguing to note that Peter Khoury claimed he had reactions—coughing and "burning sensations"—when he ingested the "alien flesh," ostensibly the left nipple of the strange blonde he claims to have bitten. His reactions were not nearly as severe as Credo's, but the African material had a less certain origin and was certainly not as "fresh" as Peter's either!

And, after the rash went down—while I was still peeling and we were smeared from head to foot with coconut oil by the initiates, every day—a strange change came over us, sir, which I am asking all people of knowledge who would read this in your country to try and explain to me. We went crazy, sir, utterly crazy.

We started laughing like real loony tunes. It was ha-ha-ha-ha-ha-ha, day after day—for the slightest things we started laughing our heads off, for hours, until you were nearly exhausted.

And then the laughing went away; and then a strange thing happened, a thing which my friend said was the goal which those who ate the flesh of a Mantindane *wanted to achieve.*

It was as if we had ingested a strange substance, a drug, a drug like no other on this Earth. Suddenly, our feelings were heightened.

When you drank water, it was as if you had drunk a wine of some kind. Water became as delicious as a man-made drink. Food began to taste amazingly. Every feeling was heightened, and it's indescribable— it was as if I was one with the very heart of the universe. I cannot describe it any other way.

And this feeling of amazing intensity of feeling lasted for over 2 months. When I listened to music, it was as if there was music behind the music, behind the music. When I painted pictures—which is what I do for a living—and when I was holding a particular color on the tip of my brush, it was as if there were other colors in that color. It was an indescribable thing, sir. Even now I cannot describe it.

The Mantindane are not the only alien beings that we Africans have seen and know about, and have got stories to tell about.

Many, many, many centuries ago, before the first White-man came to Africa, we African people encountered a race of alien beings which looked exactly like the European White-man who were going to invade Africa in our future.

These alien creatures are tall. Some of them are rather well built, like athletes, and they have slightly slanting blue eyes and high cheekbones. And they have got golden hair, and they look exactly like the Europeans of today, with one exception: their fingers are beautifully made, long and like those of musicians and artists.

Now, these creatures came to Africa out of the sky, in craft which looked like the boomerang of the Australian people. Now, when one of these craft comes down to land, it creates a whirlwind of dust, which makes a very large sound indeed, like that of a tornado. In the language of some African tribes, a whirlwind is zungar-uzungo.

Now, our people gave several names to these White-skinned aliens. They called them Wazungu, a word which loosely means "god" but literally means "people of the dust-devil or the whirlwind."

And, our people were familiar with these Wazungu from the start. They saw them, and they saw that some—in fact, many—of these Wazungu carry what appears to be a sphere made of crystal or glass, a sphere which they always playfully bounce like a ball in their hands. And when a force of warriors tries to capture a Wazungu, the Wazungu throws this ball into the air, catches it in his hands, and then disappears.

But, some Wazungu were captured by Africans in the past and forcibly kept prisoner in the villages of chiefs, and in the caves of shamans. The person who had captured the Muzungu, as he is called *in singular, had to make sure that he kept the glass-globe well-hidden from the* Wazungu. *So long as he kept the globe hostage, the* Muzungu *could not escape.*

Credo Mutwa appears to be describing the Nordic type beings that have been extensively reported in the UFO literature. These "Nordics," for example, make an appearance in the abduction accounts of Peter Khoury in 1992 and Travis Walton in 1975. But they are most conspicuous in contactee accounts, in which people claim they have largely benevolent interactions with humanlike, "Nordic" aliens. Timothy Good's book *Alien Base*—an attempt to rehabilitate the dubious world of flying saucer contactees—cites one interesting example: Ludwig Pallman's contacts with "Nordic" types from the improbably named world of Itibi Ra II.[96] Other situations fall within the blurred and ambiguous continuum between contact and abduction, like the striking 1970 close encounter of the Imjarvi (Finland) skiers Aarno Heinonen and Esko Vijo. This bizarre case featured a "Nordic" female, complete with the frequently reported hand "ball" artifact and suggestions of controlling "implants."[97] Another striking example occurred during 1968 at Villa Carlos Paz in Argentina, where a "Nordic" entity with a "ball" manipulating "solid light" (coherent beams of light, ostensibly concentrated light) confronted the witness Maria Elodia Pretzel, causing her to slowly fall backwards, then rise again, at least twice, in a manner like "the swinging motion of those sorts of dolls which invariably revert to a vertical position because their spherical base is weighted."[98]

And when Africans saw the real Europeans, the White men from Europe, they transferred to them the name Wazungu. *Before we met the people from Europe, we Africans, we had met White-skinned* Wazungu, *and we transferred the name* Wazungu *to the real Europeans, from the aliens.*

The extent and variety of Credo Mutwa's alien tales seem rather troubling for the credibility of his claims. Most seem to write Credo Mutwa's tales off as aberrations or just plain fabrication. But Credo Mutwa's tales may well be something else entirely.

Credo Mutwa Considered

What are we to make of Credo Mutwa's "alien" stories?

Credo Mutwa's stories emerge from a culture marked by a distrust of whites, a distrust of authorities, a strong focus on oral histories rather than written history, and an awareness of taboos and forbidden or hidden knowledge, known to the few initiated shamans, but experienced by the many—men and women all over Africa. But "oral material" like Credo's is very difficult to judge. Is he reaching out to some extent, trying to create a bridge between cultures? Such attempts often lead to a diffusion of worldviews, either deliberately or inadvertently. Clearly with Credo Mutwa we have a person who has been widely exposed to Western influences. It would be natural for a storyteller to incorporate such material into his repertoire if it supported his or her own experiences and knowledge. For the investigator, the difficulties emerge when such influences dilute, distort, or completely change original material.[99]

Credo Mutwa may represent a particularly intriguing case study of the emergence of an "alien" story from an indigenous source. Credo claims he has been talking about "aliens" for a long time. But could his story be growing with each retelling, fueled by the obsessions and interests of those who seek him out? Or is it a naturally unfolding progression, influenced by tribal traditions and taboos? Perhaps it's like the onion whose layers are slowly peeled back to reveal layer after layer of an extraordinary mythic view.

Perhaps the "alien knowledge" is "forbidden knowledge," and would remain so except for the initiated, unless "someone in the know breaks ranks and tells all." This is what Credo Mutwa claims he is doing. Back in the sixties he was criticized by the Zulus and whites alike for revealing too much of Zulu oral traditions in his books. His epic repository of Zulu cultural knowledge, *Indaba My Children,* and his book *Let Not My Country Die* contain nothing of an explicit alien nature. The lore with an ET spin was allegedly "hidden" within forbidden knowledge circles, and Mutwa claims he had to pass some dreadful hurdles to get access to this information. Although this seems suspect to Western sensibilities, it is actually the way taboo knowledge unfolds. But how would one confirm this?

It is not easy to establish the evolution of Mutwa's alien stories. It is difficult to pinpoint who said what when, and what influences were operating at the time. Are Mutwa's stories total fabrication, contamination, some combination of distorted truth and fiction, or a fractured reflection of some other reality? It is hard enough to establish the reality of modern Western abduction stories. In cases I've investigated I always try to establish who such people have talked to, who has previously investigated or researched their account, and for what purpose, in order to determine what sort of "baggage" the abductee may have added to the story. Some stories are so hopelessly contaminated that they're just too difficult to unravel. I don't know if Credo Mutwa's tales are in this category.

Mutwa seems to have had a preoccupation with alien tales at least as early as the late sixties. When Brian (Baruch) Crowley met Credo Mutwa in 1967, they talked more about "spiritual contact" than anything else, but Crowley does remember

Mutwa mentioning something about ETs, a subject that was not "of prime interest" to Crowley at the time. Crowley did eventually go on to record some of Mutwa's "ET" Martian origin tales, apparently during the early seventies, and finally mentions them in his books *The Face on Mars* and *Return to Mars*. These tell of Mutwa's alien origin story, with the Zulus originating on Mars. The latter book includes Jean-Claude Koven's photos of Credo Mutwa's 1983 painting, which shows the *Mutende-ya-ngenge* (the *Mantindane* or "grays"), the *Muhondoruka,* and the *Muonjina.* So Mutwa was painting them back in 1983 long before John Mack, Stephen Larsen, and others began recording his stories.

Given Credo Mutwas's legacy as a legitimate teller of native stories, one has to wonder why he would want to dilute that reputation by telling tales about aliens. Could it be that, given the opportunity to travel internationally, his story-telling has been fired up again because of the stories he has heard in developed countries, and from his perspective he has seen how many such stories have intersected with his own experiences and African traditions? Once again into the fray, he is telling anyone who will listen that Africa too has these sorts of experiences and traditions, often experienced and interacted with, on a level not often seen in the West.

His story is either a remarkable testimony of the worldwide nature of the alien abduction experience, or the legacy of a master storyteller who knows how to shape his tales to serve a purpose, cultivate an audience, or further an interest. Credo Mutwa is such a complex man that it is possible that the answer lies somewhere in a broad spectrum, or on a constantly moving stage. His rather complex job description ensures that he is not easily categorized, and adding to this the complexities of the history and culture he has lived through, further ensures that he is variously celebrated, accepted, rejected, or vilified, often in direct or indirect proportion to the realities of the circumstances. Suffice to say, he resists simplistic dismissal, despite efforts to do so.

In Conversation

I spoke with Credo Mutwa in 2002 to address some of these issues.

BC: I gather most of this information about things like this happening in Africa, they were recorded in an oral way, they weren't written down.

CM: *They were handed down, but we kept quiet about them.*

BC: Why was that?

CM: *Because, sir, if you are dealing with extra super human entities, you don't know what they will do to you if you start spreading the story around . . . and an urge to avoid retraumatizing yourself kept our mouths quiet. We knew about this thing. We put it down to gods, and the less said about it the better.*

BC: As far as yourself is concerned when did you first start talking about your own experiences and UFO experiences in Africa?

CM: *It was thereabouts, sir, in 1959 and afterwards, when I was looking for help from other shamans, trying to understand clearly what had happened to me.*

BC: So that was mainly between yourself and other shamans?

CM: *Other shamans and relatives. Some of my relatives I did not even tell. For example my late wife, I never told her about this experience even though I was already married to her. I only told her about it when she too nearly had an experience like mine.*

BC: What sort of experience did she have?

CM: *She was out in her yard. She was working as a house servant for a white family and there was no one at home. She went out to bring in some clothes. . . . There above the white man's house she saw what she thought was a bus with many windows. A light shone upon her, and then she saw what she thought were garage mechanics in silvery overalls coming towards her. She ran away and closed the white peoples' door. Afterwards she was so horrified, scared, and embarrassed by the whole happening, she thought I wouldn't believe her.*

BC: When did that happen to her?

CM: *This happened to her, sir, it was around about 1962—a terrible year in South Africa. . . . Because I could see that my wife was sick, I said please trust me, Cecilia, tell me what happened! Have you been raped, have you been assaulted? She said no. . . . In the end she broke down and told me. Then I told her my own experience, and in this way I was able to help her out of her problem.*

BC: When did you start to speak more widely to mixed audiences, both black and white, about these experiences?

CM: *The first person I went to was Elizabeth Klarer.[100] I went to her for help, trying to find reasoning behind these extremely illogical things. Why was I like this?*

BC: Did you tell her about your own experiences?

CM: *Yes, sir, I did.*

BC: What was her reaction to your experience?

CM: *She said that this is a very common thing. I expected that it was common from the side of black people, but this was the first time I had heard a white person actually accepting what had happened to me without calling me a lunatic.*

BC: When did you tell her?

CM: *I'm no longer sure, but it was also again in the early sixties.*

Credo Mutwa told me he did not know Cynthia Hind, an African UFO researcher and author of two books on the subject.[101] He explained:

In South Africa for many years black people and white people did not openly talk to each other, especially on subjects of this nature. When black people try to reach out to white people, they are often rebuffed and their experiences are put down to stupid black personages. . . . The white people simply thought these were just superstitious savages. Everything was put down to black superstition, and it still is. . . . Africans simply tell what they know. . . . These creatures have been known by human beings for millennia. Africa is the unexplored country, the neglected country, the ignored country, and yet it is a golden link that joins all human experience throughout the world together.

There are things that [Western writers] refuse to know simply because those things come from Africa, and yet I have traveled to many parts of the world. I have found these amazing things between every country . . . [in common] with Africa. . . . The world is one, and the vastness of human experience in many possible fields is one.

BC: I noticed when you described your experience of 1959, I think in one of the accounts that I've heard you talk about, you mentioned the strange lady you saw inside that strange room, the blond hair. . . . There was some mention she had unusual hair coloring in other parts of the body as well?

CM: *You mean the creature that assaulted me, the female creature?*

BC: That's correct.

CM: *This creature was strangely like a white person, but I am a painter, a wood carver, and a sculptor. I can tell you that the body of that creature was out of proportion. The breasts were set too high in the chest, and the legs were too short for the rest of the body. The eyes were slanting, very, very bright blue, almost as if they had electric lights behind them. The hair was golden—as if the hair was made of metal or some nylon substance. . . . It looked too metal-like . . . if you can image fine golden fiber. It was arranged like a bun at the back. . . . The hair was split in the middle. . . .*

BC: What about other body hair?

CM: *The hair under the armpits and between the legs was blood red, very red against the . . . very pale skin.*

BC: How would you describe her skin coloring?

CM: *It was very, very blond, very fair skin, with a kind of luminescence to it, as if it was polished.*

I asked Credo Mutwa about his awareness of the Antonio Villas-Boas case that occurred in Brazil in 1957 and his claimed encounter with a blond woman, which I told him was very similar to his own description. He brushed the question aside, indicating:

CM: *The thing is, sir, I am not the only one. There were others, black men and Hindu, who had similar experiences like mine here in South Africa.*

When I said it was intriguing that the descriptions were similar even down to the blood-red body hair, he responded:

CM: *Like I say, sir, creatures of this kind have been known in Africa for millennia. We have a name for them—Ncupusana, which means the trappers, the one who entraps people. . . . These blond people, our people have seen them, females as well as males . . . Many behave in that way* [they did to Credo Mutwa] *in conjunction with the gray creatures. Some of them just come and are seen in villages, and sometimes they try to leave [strange] messages to people, messages which make no sense to the African people . . . for example, "In the vastness of the cosmos there is a thinking brain." Now what is an African to make of that kind of gibberish . . . ?*

BC: A lot of the Western researchers feel they were the ones who discovered this phenomenon, in the sense of the abduction phenomenon.

CM: *Africa has been totally ignored by many arms of research in the Western world, and people like me can't correct this great injustice and imbalance. There is more to learn in Africa than in any other continent on earth.*

I asked him what his own country people's reaction had been to his extraterrestrial experiences.

CM: *People are benighted here. People are asleep here. In many parts of South Africa people are still caught in the Victorian mindset. I'm referring to the white people, and black people are caught in the good spirit, bad spirit phenomenon, and they explain everything in terms of spirits. . . .*

I asked Credo Mutwa about comments attributed to him in which he suggested that maybe these beings were not extraterrestrial but were from our future.

CM: *I think so, sir, from what I have seen and what little I have found. For example, a creature from outer space would walk into our earth with its entire body protected by some kind of armor. But there are so-called extraterrestrials which walk about with their faces bare. . . . The pink-skinned aliens, they go about with their faces exposed. They cannot be extraterrestrial creatures. . . . If you go to the planet Mars, being an Earth man, you would have to wear a helmet to breathe with. But here we have an interesting phenomenon, which I have been studying for years, of creatures which walk about with their hands unprotected and their faces unprotected, which means they breathe the same air we breathe, and for some reason are not afraid of the many microbes which could attack them. . . .*

Perhaps the startling DNA work done with Peter Khoury's "alien" hair sample may hint at a reason for this behavior, namely, that "aliens" may have genetically engineered an immunity to some, if not all, human diseases. For example, the owner of the blond hair sample may have had the genetic factor of CCR5 deletion that seems linked to a resistance to things like AIDS.

Credo Mutwa basically argues that the picture is not a simple one and may involve not only extraterrestrials, but time travelers, as well as an advanced civilization in the past.

We discussed the Jean-Claude Koven photos of Credo's 1983 alien paintings. Credo Mutwa indicated he had been painting these sorts of alien paintings since he was a child. He also lamented that all his paintings were stolen. But he rationalizes laconically:

CM: *It's life, sir.*

Now what I do to heal people, I've also made little sculptures out of aluminum, which I melt in the African way, and I've cast different kinds of star creatures, so that when a person comes to me, who appears to have undergone a traumatic experience like that, I don't cross-question them. I ask them to go into a room and to pick out the one creature which was involved in his or her experience. Scores of people in the last thirty years or more always pick out the Mantindane. . . . *You see, people come to me and other* sangomas *in search of help when they have undergone such an experience. We are the only people they can talk to.*

Despite this, there is still a dark legacy of bitterness and hatred.

CM: *Some black people hate me bitterly for telling white people about these things. They say we should be quiet . . . because of [the] breaking [of] a taboo, it is associated with deep guilt.*

Credo indicated there are taboos about talking about these alien beings, that he had broken those taboos and many others, and stated his reason for doing so:

CM: *To help my people, because our people do not know their true greatness. Our people have no feeling of their self-worth as a people. They know a lot of things and yet they are content to say they know nothing, and that it is the white man who knows everything. Yet this is utterly false. [The white people] don't show arrogance because they are evil people. No one is evil in this world. It is because they don't know. . . . Ignorance is the reason why we have got so much violence and bloodshed on the earth. . . .*

He reflected that people like himself who have had these experiences develop a kind of faith based on an expanded awareness of reality, beyond the shallow reality of our daily lives. However, his own circumstances had changed. He indicated he was no longer going to speak out about these things.

CM: *When my wife died after forty-six years of marriage, something has died inside me. I'm beginning to say, what's the use of it all? . . . What is the use of talking where there is no one to hear? . . .*

I suggested that a lot of people are listening.

CM: *Yes, sir, but not the people that matter. . . . The biggest danger is ignorance. . . .*

People who have really been abducted, people who carry the strange stigmata on their bodies develop a compassion for humanity, which is almost pathetic, and I am like that. I loathe and detest myself for what I feel for people. I don't know why I feel this way, but all those who went through the same horror I went through develop a love for humanity that is almost insane. When you hear about war you have a feeling deep in you like that of a woman about to give childbirth. Believe me. . . .

All one can do, who have gone through the hands of the so-called gods, is to sit in a dark place and cry, and ask yourself, don't these fools know we are not alone on this earth? Why can't they see what's going on around them and do something about it? Humanity is being the lamb of sacrifice to entities that they don't even believe exist. . . . There are gray aliens! Illusions and dreams and hysteria do not leave stigmata upon the bodies of hundreds of men and women in remote parts of the world. No!

Credo Mutwa is frustrated and embittered by the legion of obstacles people like him and his own people have had to endure. He maintains a conspiratorial view of who really is controlling the world. His is a bleak and unrelenting worldview. He expresses a kind of resignation and a feisty acceptance of his lot and the life he has had—as a writer of eight books, for which he claims he gets no royalties, despite the fact that they continue to sell; as an artist since 1935, with extraordinary paintings, giant concrete statues, and many other works to his credit, though he remains unrecognized as an artist in his own country; as a healer with cures for major diseases. Despite this legacy[102] and much more, he argues there are forces and people who prevent the fuller expression and use of these assets. His frustration is palpable.

CM: *I am glad my life has come to an end. I don't know why I have lived for so many years. Eighty years! Going for eighty-one. It's just too much....*

A Shamanic Perspective

Credo Mutwa's story offers us a chance to reconsider the shamanic perspective on such extraordinary experiences. At the turn of the last century, anthropologists Walter Spencer and Francis Gillen, in their book *The Northern Tribes of Central Australia,* provided a classic account of the remarkable shamanic tradition. An aborigine, Kurkutji, was set upon by two spirits, Mundadji and Munkaninji, in a cave:

"Mundadji cut him open, right down the middle line, took out all of his insides and exchanged them for those of himself, which he placed in the body of Kurkutji. At the same time he put a number of sacred stones in his body. After it was all over, the youngest spirit, Munkaninji, came up and restored him to life, told him that he was now a medicine-man and showed him how to extract bones and other forms of evil magic out of them. Then he took him away up into the sky and brought him down to earth close to his own camp, where he heard the natives mourning for him, thinking that he was dead. For a long time he remained in a more or less dazed condition, but gradually he recovered and the natives knew that he had been made into a medicine man. When he operates the spirit Munkaninji is supposed to be near at hand watching him, unseen of course by ordinary people."

This is an excellent description of the initiatory experience of an Australian aboriginal shaman. Ritual death and resurrection, abduction by powerful beings, ritual disembowelment, implanting of artifacts, aerial ascents and journeys into strange realms, alien tutelage and enlightenment, personal empowerment, and transformation—these and many other phenomena are recurring elements of the extraordinary shamanic tradition.[103] Obviously, these worldwide native aboriginal experiences share many impressive similarities with the bizarre complex of human experiences now holding sway on the world stage, namely alien abduction encounters.

Whitley Strieber, in trying to come to terms with his own abduction experiences, wondered "if the shamanic language of symbol and myth would offer a better insight into the visitors' motives." In an interview for *Terror Australis* magazine, Strieber stated that "shamanism is the shattered remnants of mankind's early attempts to control this [visitor] phenomenon."

In shamanism we see individuals who seem to have a strong measure of control over the set of realities they operate in. But in UFO abduction experiences we appear to have generally helpless victims who have no control over the bizarre "reality" that overwhelms them. The two types of experiences appear to be on opposite ends of a control continuum. Perhaps we as a culture have lost an ability that other cultures and generations may have had to some extent. Are we now undergoing "rites of passage" that will enable us to achieve some level of understanding and control over the UFO abduction experience?

Peter Khoury, the Australian alien abductee, in 1996, whose 1992 experience yielded the hair sample and DNA analyses featured in this book. (CREDIT: PETER KHOURY)

This scoop mark appeared on Peter Khoury's leg from his 1988 experience. Such marks appear frequently in abduction cases. If the biochemical nature of the rapid healing often reported can be established, the benefits in human health could be enormous. (CREDIT: PETER KHOURY)

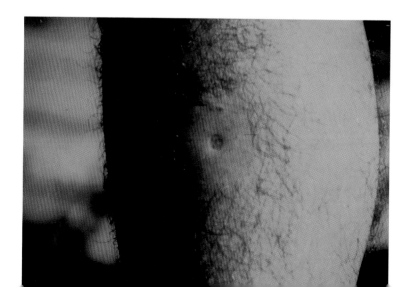

This painting captures elements of his frightening 1988 encounter with the two types of entities and the needle-like device being inserted into his head.

(CREDIT: PETER KHOURY WITH CHRISTINE HILLMAN)

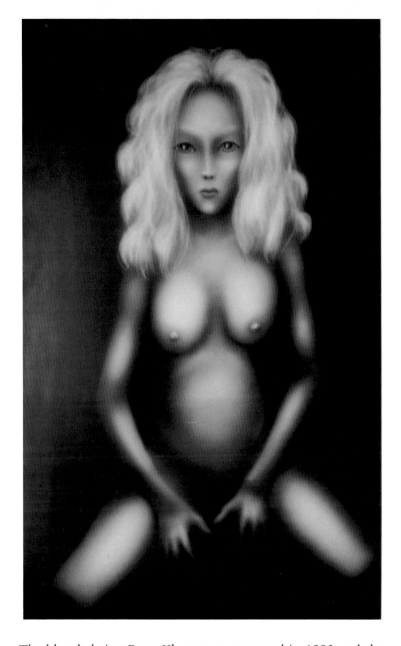

The blonde being Peter Khoury encountered in 1992 and the apparent source of the strange blond hair sample which yielded extraordinary DNA data (CREDIT: PETER KHOURY.)

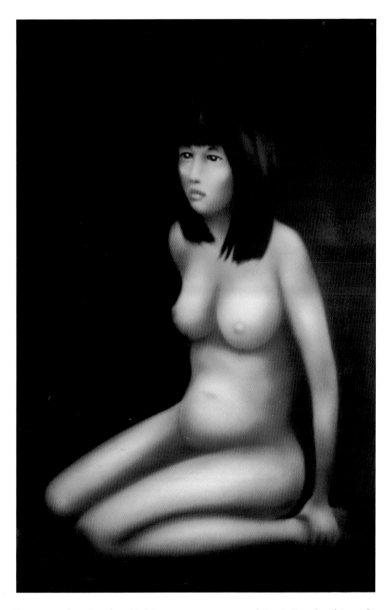

Present also in the 1992 encounter was this Asian-looking female being, who seemed to be observing the bizarre interaction between Peter Khoury and the blonde being. (CREDIT: PETER KHOURY)

This is the DNA sequence extracted from the shaft of the 1992 blond hair sample, which revealed very rare and unusual Asian mongoloid DNA results.

(CREDIT: APEG/BILL CHALKER)

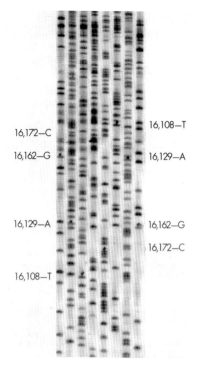

16,172—C
16,162—G
16,129—A
16,108—T

16,108—T
16,129—A
16,162—G
16,172—C

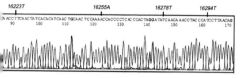

This is the DNA sequence extracted from the root of the 1992 blond hair sample which revealed very unusual hybrid-like DNA sequences indicating both rare and unusual Basque/Gaelic and Asian DNA results.

(CREDIT: APEG/BILL CHALKER)

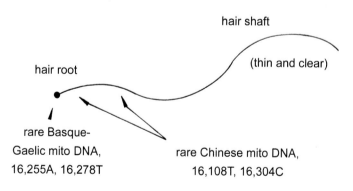

hair shaft

(thin and clear)

hair root

rare Basque-
Gaelic mito DNA,
16,255A, 16,278T

rare Chinese mito DNA,
16,108T, 16,304C

This drawing summaries the unusual DNA results found in the hair root and shaft areas—results suggestive of advanced cloning techniques and possible hybrid characteristics.

(CREDIT: APEG/BILL CHALKER)

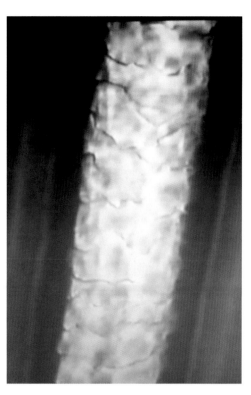

The alien hair sample revealed under a powerful microscope examination.
(CREDIT: APEG/BILL CHALKER)

The DNA results here from the hair sample are suggestive of CCR5 gene deletion—the extraordinary factor implicated in HIV and other virus resistance.
(CREDIT: APEG/BILL CHALKER)

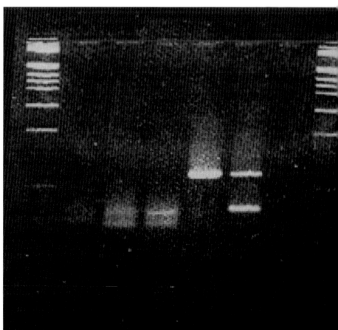

Vicki Klein at Dalton—she was the focus of an intriguing alien journey between 1969 and 1972 that prefigured the current idea that abductions involve genetic experimentation.
(CREDIT: MARION LEIBA)

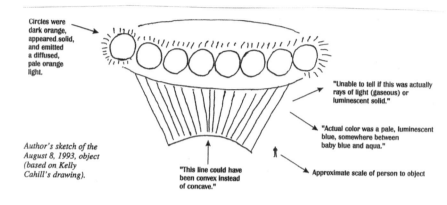

Circles were dark orange, appeared solid, and emitted a diffused, pale orange light.

"Unable to tell if this was actually rays of light (gaseous) or luminescent solid."

"Actual color was a pale, luminescent blue, somewhere between baby blue and aqua."

Author's sketch of the August 8, 1993, object (based on Kelly Cahill's drawing).

"This line could have been convex instead of concave."

Approximate scale of person to object

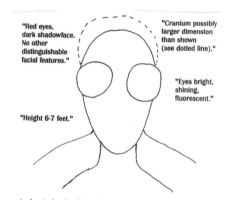

"Red eyes, dark shadowface. No other distinguishable facial features."

"Cranium possibly larger dimension than shown (see dotted line)."

"Eyes bright, shining, fluorescent."

"Height 6-7 feet."

Author's sketch of being (based on drawing by Kelly Cahill).

The UFO and entity in Kelly Cahill's 1993 encounter—an extraordinary case for the reality of alien abduction.
(CREDIT: INTERNATIONAL UFO REPORTER (IUR) SEPTEMBER/OCTOBER 1994 BASED ON AUTHOR'S RECREATIONS OF ORIGINAL SKETCHES BY KELLY CAHILL)

A comparison of accounts of alien abductions and shamanic initiations is revealing.

Folklorist Dr. T. E. Bullard lists the following elements of UFO abduction narratives:

- capture
- examination (which includes tales of specimen taking, reproductive examinations, implanting of small artifacts inside the abductees, and removal of organs)
- conference (the witness talks with abductors)
- tour
- otherworldly journeys
- theophany (the witness has a religious experience or receives a message from a divine being)
- return

Michael Harner, Holger Kalweit, and others list the following recurring elements of shamanic experiences:

- selection or capture
- initiation experiences (which include accounts of implanting artifacts, removal of organs, etc.)
- "magical flights," including "celestial ascents" to strange realms
- dialogues with mythic personages or spirits
- return

Both types of experiences involve the placement of artifacts within the body. In shamanism it is crystals and the like. With UFO abductions, it is "implants." Celestial ascents are also common to both genres. Even the descents into "other realms"—the underworld or undersea world—that occur in shaman stories, occur occasionally in UFO abduction stories. And animals often feature in both experiences. In the case of shamanism, deer, birds, fish, wolves, bears, and other animals take the form of "power animals" or "spirit allies." In UFO abduction accounts, the presence of animals is rationalized as "screen memories" of aliens. In his book *Missing Time,* Budd Hopkins describes Virginia Horton's encounter in a French wood with a talking deer that turns out to be an alien. And recall biochemist Kary Mullis's strange encounter with a talking, glowing raccoon.

The sexual/reproductive element also appears in both shaman and UFO abduction narratives. In shaman accounts we even have stories of "spirit weddings" and "spirit children" offspring—the results of liaisons between shamans and the "spirits." Holger Kalweit gives a fascinating account of such an encounter in his book *Dreamtime and Inner Space.* The shaman "gave birth to a child by [a spirit] in the spirit world, and he would bring it to her at night for her to breastfeed it. He came when everyone was asleep. The people in the village heard the child cry, but her own family slept as if they were dead." [104] It is fascinating to compare that sort of story with the accounts Budd Hopkins describes in his book *Intruders*— tales of ova sampling, baby presentation, and baby bonding— as well as Vicki Klein's story.

Some of the similarities between the two types of encounters argue for a literal extraterrestrial interpretation of such stories, but one should be careful not to uproot such tales from their cultural setting and to stamp them with Western cultural imperatives, such as our fascination with extraterrestrials and their possible relevance to UFO experiences. This caution is particularly relevant to tribal legends and oral traditions. Even though similarities occur between UFO abduction stories and the shamanic experiences of aboriginals, a literal interpretation may be incorrect. Instead of extraterrestrials we may be dealing with something else. The perspectives offered by indigenous cultures such as that of Australian aborigines may support other interesting possibilities. Such experiences may be about the effect of subtle forces exerted on humans by the natural landscape,[105] "places of power" (or locations of unusual natural energies),[106] hallucinogens,[107] our modern electromagnetic environment,[108] or so-called "earth lights."[109] The popular acceptance of a literal extraterrestrial explanation as an answer may be premature or incorrect.

I spoke to a number of people about this perspective, including Robert Lawlor, author of *Voices of the First Day: Awakening in the Aboriginal Dreamtime*,[110] who suggested that the ET perspective plays into the vanities of the evolutionary worldview of the Western world. The indigenous view of the shamanic world validates the realities and the cultural development of people in harmony with nature. The former is more palatable in the Western technological setting. Lawlor suggests that the rash of modern UFO abduction experiences may in part be due to the effects of an increase of industrial energies or electromagnetic pollution on our neural perception modalities, and the indigenous accounts may be due to abnormalities in the electromagnetic ambience of the natural landscape, along with increased sensitivity due to initiation procedures and ceremonies. There are obviously many perspectives to examine when considering UFO experiences beyond the obvious mainstream appeal of the extraterrestrial hypothesis.[111]

Credo Redux

In the *Profiles of Healing* book edited by Bradford Keeney, Credo Mutwa recounts an intriguing anecdote. He describes how in 1918 his mother mysteriously disappeared from her community for a number of days.

No one believed her story about the disappearance, and she was thrashed with a hippo-hide whip for it. She said that she was taken away by a little old man, with large eyes, who showed her many strange things. My mother, when she came back, carried a peculiar smell, and the wise people in our community told my grandfather to believe her because she carried the smell of a Mantindane. *This is a Zulu word too often incorrectly translated to mean a fairy or a nature spirit. More accurately, it refers to what some people call the gray aliens. After I was born [in 1921], my mother and I both disappeared and came back a few days later. This time she returned completely gray, looking like someone who had died, and everyone believed her story.*"[112] Credo Mutwa contends these sorts of encounters with unknown entities are common in Africa.

Credo Mutwa's stories of personal experience with alien abductions, while suggestive and perhaps compelling, fall short as entirely credible evidence as they are part of his early origin stories, which started surfacing only in the late 1980s and early 1990s, consistent with the Western dissemination of such stories. There seems more compelling evidence that Credo Mutwa was talking more generally about aliens and UFOs in the 1960s and 1970s. Either Western abduction stories inspired Credo Mutwa's stories, or they are authentic, but somewhat colored by the processes from which they emerged.

CHAPTER ELEVEN

Hidden Evidence

THE ABDUCTION PHENOMENON IS AN EXTRAORDINARILY COMplex human drama, but to date it has not provided the kind of physical evidence that would persuade a scientist. UFO and abduction encounters are marginalized experiences in human society. Essentially they are seen as fringe phenomena. They are not accepted as reality. We live in a physical reality, and that's why there is an emphasis on the search for physical evidence. In our time and our culture it's science that decides what is real or not. One of the best ways to obtain physical evidence is to treat the scene of an alleged alien abduction as a "crime scene." This approach has now been advocated for about a decade, and as we have seen with the Peter Khoury case, has begun to reap dividends. While DNA profiling dominates the public perception of forensic investigations, it is only a part of a much larger toolkit.[113]

Crime Scene

In 1993 researcher Victoria Alexander issued a call to arms for a "new protocol for abduction research" using the techniques of criminal investigations. She was responding to what she saw as the passive collection of abduction stories and suggested that abductees could contribute more actively in collecting potentially critical data. Alexander recommended the formation of an Abductee Task Force (ATF) that would utilize Abductee Crime Scene Unit protocols. She elaborated with suggestions that follow standard crime scene forensic protocols: "sealing" the scene, photographing the site, identifying and sampling entry and discovery points, and utilizing sampling kits. Alexander suggested that "police rape kits" could be used, with abductees themselves collecting the more "intrusive" samples, such as swabs from pubic areas and vaginal washings. Clothing touched by "aliens" might also provide vital evidence. Surfaces that could hold fingerprints, such as human skin, painted surfaces, glass, etc., might yield "extraterrestrial fingerprints." In a variation on "thief detection powder" and stains, Alexander suggested coating bedroom floors with flour in the hope that aliens might walk over them and perhaps leave a "footprint." She even suggested that traces like "hair or bodily secretion might be secured from the alleged 'hybrid' offspring of alien-human genetic experiments." Alexander bemoaned all the "missed opportunities," like the case of New Jersey abductee and artist David Huggins, who claims repeated sexual encounters between 1963 and 1987 with an alien female who had "head and pubic hair which seemed unnaturally perfect, as if wiglike, and very black in color."[114] As far as I know, Huggins has never obtained any such hair samples for analysis, nor am I aware of any significant results by Alexander in carrying out any of her forensic suggestions.[115]

Other researchers have also picked up on the forensic approach to alien abductions.[116] One is Derrel "The UFO Hunter" Sims, if one believes the material that emerged from his Houston, Texas–based UFO abduction group during 1992 and 1993. Sims tells stories of "forensic" hypnosis, abductees inserted into abductions as "Manchurian candidates," mass abductions, and recovered "alien implants," but the evidence he offers is mixed, ambiguous, and ultimately unconvincing.[117]

Another researcher to take the proactive forensic approach is Cheryl Powell, whose background is in nursing and laboratory medicine. During 1995 and 1996 Powell and her Goffstown, New Hampshire, group, called the International Center of Aerial and Abduction Research, tried to apply a forensic approach to their abduction research. They reported nothing of significance, however, beyond finding some very small footprints in relation to an abduction case and a possible "Bigfoot" hair sample. Powell's laboratory medicine background also led her to suggest that the blood chemistry of abductees should be checked.[118] New Jersey investigator George Filer also suggested this sort of approach, pointing out that "analysis can determine if they have been flying at high altitudes in the recent past." NASA's biochemistry studies of astronauts provide potentially interesting insights into the effects of altitude and weightlessness on a person's blood and other bodily functions.[119]

Yet another researcher has investigated a forty-year-old dress, looking for clues to back up a claimed alien encounter. Late in 2003 chemist Phyllis Budinger circulated her analysis of the large pink stain found on the dress Betty Hill wore during her famous abduction of 1961. The dress had been hanging in the closet of her New Hampshire home relatively undisturbed all those years. Could the stain and possible residues be related to her handling by the alien beings she described during her encounter? In her analysis, Budinger found the presence of amides in the stain, which suggests the presence of proteins. This made the search for possible DNA residues worthwhile. My APEG team received some dress samples and control biological samples pertinent to Betty and Barney Hill. A specialized (ribosomal) PCR DNA profiling was undertaken to extract non-human DNA along with standard assessments. While our results revealed nothing of an explicit alien nature, they confirmed the usefulness of the techniques. Despite the passage of decades, DNA found on the dress was connected to Betty and Barney Hill, and rather surprisingly, at least some of the stained areas were found to be due to spider's blood. At best, these and other DNA results may support Betty's claim of being in a wooded setting during their abduction. Sadly, Betty passed away before this work was completed in December of 2004.[120]

Alien Dust

A biophysicist better known for his controversial testing of crop circles[121] has also gotten in on the act. In 1997 Dr. William Levengood of Michigan was contacted by an abductee who described being woken up the previous night by the presence of a beam of light that shone onto the floor next to her bed. She claimed that when she put her arm into the beam, the beam shimmered. According to Marilyn Ruben, who runs the Alien Abduction Experience and Research (AAER) website, the woman then experienced a nocturnal encounter with aliens. Levengood visited the woman's home and took samples of "the whitish dust residue on her furniture where the [witness] had reported seeing the sparkling beam of light during the night." Levengood, who runs his analytical work through Pinelandia Biophysics Laboratory, reported unexplainable "glassy particles" and "pseudo crystals" in the dust samples.

Further studies of the household dust of other abductees finally led Levengood to announce his findings in February 2000. Suddenly ordinary house dust was coming under scrutiny for possible evidence of alien abductions. Microscopic "glassy particles" and "pseudo crystals" were being touted as evidence of alien intrusions. Soon the AAER website was asking: "What is your house dust I.Q.? Has your home been visited by aliens? Get your house dust tested and find out!" Their ad gives basic instructions for sampling with a cotton swab, placing each sample into separate plastic bags identified with name, room, and date, and then sending them to AAER "with your check or credit card information." Lots of pretty photographs of "pseudo crystals" have been posted on the AAER website, but given the huge variety of particles that can get incorporated into household dust, and the extreme difficulties in establishing relevant and comprehensive control samples, I suspect this exercise of identifying "pseudo crystals" will do little to contribute to our knowledge of the reality behind alien abductions.[122]

During September 2000 Australian researcher Barry Taylor released details of DNA testing on a paw print left on a mirror in a residence on the north coast of New South Wales. He felt that the "prints" may have been digit prints and could be alien abduction related, but he has provided few details to justify such a conclusion. The results revealed canine DNA and the presence of "an unusually high density bacterial culture," which may be encountered in soil and water. While I encourage the application of DNA testing, I suggested that some effort be made to do control DNA testing on possible canine candidates, but these controls do not seem to have been carried out, and therefore the results remain inconclusive.[123]

Implants

The strange artifacts or implants that abductees, such as Peter Khoury and Whitley Streiber, have been reporting in their bodies following their experiences are another potential source of evidence for the physical reality of alien abductions. The problem is that the implants are rarely recovered; Peter Khoury, as we have seen, lost his. But since the mid-1990s a concerted effort has been made to recover these artifacts, largely through the efforts of California podiatric physician and surgeon Dr. Roger Leir, who has facilitated the surgical removal of a number of these strange items and has had them analyzed.

The first of his "implant" removals occurred on August 19, 1995, when medically supervised surgical procedures isolated three objects from two abductees known as "Patricia" and "Paul." In the case of Patricia, two items—a T-shaped metallic artifact and a metallic "seed"—were removed from her toe. Another metallic "seed" item was removed from Paul's hand. Since then Roger Leir has described about a dozen such surgical "implant" removals. The National Institute for Discovery Science (NIDS), which supported Leir's work, was able to conduct physical analyses of the implants through a laboratory at New Mexico Tech, an engineering university located in Socorro, New Mexico, including X-ray energy-dispersive spectroscopy and X-ray diffraction work.

A few, namely Leir's first three implant removals, suggest compositional results and isotopic ratios that lend support to the idea that they may be of extraterrestrial origin. In short, the implant materials seemed to be of meteoritic origin. While provocative, these correlations are tentative and were made on the basis of technical interpretations from meteorite data. This is a long way from intelligent extraterrestrial sources. The New Mexico Tech analysis also offered a more prosaic origin for the implants: a stone, for example, subjected to calcification because of its apparent embedding inside the body of a person. A final answer, they concluded, can come only with "more in-depth studies." [124]

One of the more interesting "implant" studies to emerge was reported at the Abduction Study Conference held at the Massachusetts Institute of Technology (MIT), during June 1992. At the conference and in a subsequent paper, MIT physicist Dr. David Pritchard described his efforts at identifying a small foreign item that emerged from the penis of a New York State abductee named Richard Price. Price had a recollection of an alien abduction experience in 1955 when he was only eight years old, during which he saw something implanted into his penis. In 1981 there was medical verification of the presence of a subcutaneous "foreign body roughly 4 mm long and 1 mm in diameter" in his penis. By 1989 this artifact was beginning to emerge from the skin surface and was verified by Dr. Pritchard at the time.

Price said that the item finally dislodged on August 10, 1989, seemingly creating a "strong electric shock" sensation as it did so. Price placed the artifact in a clean film container and showed it to Pritchard twelve days later. This kind of detail is important from a forensic point of view because it establishes a "context" for the item in terms of its possible connection to an alleged alien abduction experience. In other words, there seemed to be a verifiable "chain of evidence." Dr. Pritchard described this in terms of "pedigree," namely "what ties the artifact to the abduction phenomenon. It includes the linkage of the artifact to a particular abduction report, the similarity of the abductee's experience to those of other experiencers, the testimony concerning the artifact's origin, any documentation accompanying this testimony, the circumstances of its recovery, any prior descriptions of the artifact or descriptions of similar artifacts in independently investigated cases, the temporal duration and stability of the abductee's claims, etc."[125] This was extremely important for Dr. Pritchard, just as it was critical for me to establish such issues in the case of Peter Khoury's hair sample and the experience from which it emerged.

The study Dr. Pritchard describes revealed a probable prosaic origin for the implant and failed to confirm the existence of an extraterrestrial artifact. He concluded, "The analysis shows nothing unterrestrial . . . [with] the irregular overall characteristics of something that grew . . . [i.e., in the body]. Its elemental and chemical constituents are also consistent with earthly biological origins." In short, the implant seemed to be calcified tissue due to a piece of foreign material like glass or wood, or from trauma of some kind, perhaps a table corner.

Dr. Pritchard highlighted the cautionary nature of his investigation. Results that suggest "unusualness" need to be carefully assessed, he said. Dr. Pritchard's study was hampered by the lack of control comparisons, such as we had for our hair studies. Caution is indeed our best ally. The wide study of prosaic items and factors is needed to help establish whether any alleged evidence is truly anomalous. It is a long and difficult process, in which certainty might never be established. This is why cases like "implant" investigations, and indeed the Khoury hair sample, must be left open, subject to ongoing research and investigation.

David Pritchard also noted that all such studies need the benefit of "a multidisciplinary group with chemists, biologists, and material scientists of various sub-specialties as well as several experts on each of the various machines and techniques used in the analysis." He highlights the difficulty in doing this, because of the lack of support and recognition the UFO field has from mainstream science. Fortunately, because of his reputation and position, Dr. Pritchard was able to obtain such support, which ultimately revealed the prosaic origin of the Price "implant." [126]

If the emerging "implant" data can be confirmed via more comprehensive and multidisciplinary scientific studies, then the question that arises is: What are such items doing inside the bodies of abductees? What is their function? Some researchers believe they might serve to "track" or "monitor" abductees, but the evidence so far revealed on implants does not support such notions. Keith Basterfield recently undertook a detailed review of UFO literature accounts of "implants." Accounts seemed to have emerged from 1979 onward, even though abduction accounts occur significantly earlier. Basterfield's study [127] highlights the lack of consistency in implant descriptions and their locations in the body, and the lack of peer review for the analyzed data. So the variety of such artifacts, in both their characteristics and their placement, suggests that we are a long way from certainty about them.

On Camera

In an attempt to acquire definitive proof of the reality of alien abductions, a number of researchers and abductees have tried to capture these events with video cameras or webcams. In most cases results have been controversial and ambiguous, especially given the extraordinary things that can be done in image manipulation today. I am aware of no case to date of a credible video, webcam, or photographic image of an alien abduction that has been thoroughly investigated, well documented, and made widely available for peer review by UFO researchers and interested scientists.

I'll be the first to admit that some cases are tantalizing, however. For example, in 1999 and 2000 John Carpenter, who was then a MUFON (Mutual UFO Network) consultant, claimed that a security videotape from a Florida factory was the first to show "an abduction . . . and the worker's return one hour and 50 minutes later in a laser-like 'puddle' of light." He cited video and scientific analysis to back up his claim, but since the basic context of the case, such as who, where, and when, were never established, the case remains highly questionable. An anonymous video that is aired on television with no corroborative details falls far short of the kind of credible evidence needed as proof of the reality of alien abductions.[128]

Others have also come forward with claims of photographic evidence. California researcher Ann Druffel has presented photos, taken in Pasadena by Baptist minister Rev. Harrison Bailey in November 1978, of a metamorphosing "alien entity." The photos show an animate shadowy being as well as partial bodily aspects "hiding" behind Halloween masks.[129] Budd Hopkins has presented some curious, red-tinted photos of a playground area in suburban Brisbane in Australia during 1992 that may have some significance in a family's abduction account. Hopkins cites the photos as displaying possible evidence of the invisibility techniques employed by alien abductors.[130] In 2003 Italian researchers Roberto Malini and Federico Dezi cited an alleged South African abductee's webcam imagery of moving shadows— "creatures"—as inspiration for Operation W, their own webcam-based monitoring of abductees. They hope to capture an alien abduction in progress.[131]

Photographic evidence of this kind will go beyond curiosity only when such "proof" emerges from controlled studies. Currently, the most ambitious and promising step in that direction is the Ambient Monitoring Project (AMP), an effort begun in 1998 by the UFO Research Coalition, which consists of MUFON, the Fund for UFO Research, and the Center for UFO Studies (CUFOS). Dr. Mark Rodeghier, scientific director of the Center for UFO Studies, described the origin of the AMP project in the January 2002 issue of the *MUFON Journal:*

"Some investigators over the years have attempted to use a video camera in the bedrooms of abductees to obtain physical proof of an abduction event. These efforts, though, have met with little documented success. If abductions are, at least in some instances, physically real, then we would expect that those beings doing the abductions would leave some trace in the environment. Even if super-aliens have the ability to transfer a person through a bedroom wall to a waiting UFO, that doesn't imply that the environment in the bedroom would be unaffected while the transfer was occurring. Based on this assumption, the UFO Research Coalition decided several years ago to initiate a research project to monitor the potential physical changes associated with an abduction event.

Our working hypothesis is that some physical changes must occur during a real abduction event (although we are uncertain as to exactly what these might be).

"Specifically, we have constructed a device that is placed in the home of an abductee and left there for about 4–6 months. The device is small and attractive, about as big as a woman's makeup case, designed to blend into the home. It records various physical parameters and stores the information, which is then downloaded to the computer of our project engineer (Tom Deuley, a longtime MUFON official) every day.

"While the device is in her home, the abductee keeps a daily journal of experiences related to the abduction phenomenon. Then, when the data collection period is over, the journal will be matched up to the physical data to see what was recorded at those times when a person thinks that she may have been abducted, or something similarly odd occurred.

"The project officially began in the spring of 1998. Those most involved with the project include Rob Swiatek from the Fund, myself from CUFOS, and Tom Deuley. AMP started with the development of two prototype devices, used in three pilot cases. From this experience, we constructed the final monitoring device, which was used for the first case in the summer of 2000."

By January 2002 the AMP project had "successfully collected data from several abductees with few technical problems." The project operated as a type of "double blind" experiment, Rodeghier explained: "Our project engineer, who collects the data, has no idea what is recorded in the journal by the abductees. And the abductees and UFO investigators are not told what the device has recorded. The data will be archived and later, when the project has ended, be compared and correlated to determine what can be learned. We follow this protocol—which means that there are no results yet to report . . . —because it is crucial in the study of anomalies to design rigorous research that removes any chance to alter the data to fit some preconceived notion. As we know, there are skeptics and debunkers who are only too happy to find deficiencies, real or imagined, in any serious research project. The results of AMP, whether positive or not, will potentially be quite important for ufology. Obtaining physical confirmation of an abduction event would be an important advance in our study of UFOs. Finding no physical evidence of abduction events would be open to several interpretations, but it would be just as important a finding." [132]

In 2003 Tom Deuley gave me a status report on the AMP project: "The data collection portion of the experiment ended on June 15, 2003. It is estimated that it will take 6 to 9 months to analyze the data collected. We collected about 1.2 gigabyte [worth of data] for each of 13 cases. It will take a bit of time. As the collector, I did not see anything unusual, but it all must be bounced against the journals the subjects kept. As soon as we do the evaluation, a report will be published and made available. If there is any strong correlation, I am sure we will refine the equipment and step even higher into doing solid science. In any case we feel we have done good work. Details of the equipment and the work will not be available till after the final reports are published.

"For the moment we are not using the devices for any other work until our data has been completely analyzed and the final report made. We feel we must protect the device and its capability until that is all completed. Eventually we will publish the report and then will be glad to discuss the device at any level. I can say that after monitoring every file from sixteen cases we never saw a power failure or anomaly that we could not explain.

"The old idea that the 'aliens' would know and either turn things off or fool the device has not proven out in our work, at the same time the subjects did continue to report incidents. The individual channels were what one would expect with reasonable sensitivities with readings being logged each second 24/7. It is completely passive except during downloading, which was always done when no one was home. A new design would require considerable updating."

I asked Deuley whether any prosaic actions by the "abductees" could create a false positive with something recorded in the journals.

"Yes," Deuley replied, "they could try, but they could not understand the interactions between channels which would in general divulge the obvious try at fooling the system. With single instruments they have the advantage. With an array there are cross feeds that protect from fooling. For instance if I were to put a video camera in a room, I would add a separate but undivulged audio recorder for, let's say, comparison's sake.

"We do protect ourselves by having multiple channels and base understandings of the interactions we should see when any channel is activated. It is not 100 percent, but is unknown to the subject. We also would know if the subject fooled with the box, the power lead or the telephone lead, or if they attempted to open the device.

"Finding subjects was our biggest problem. We at first wanted 60—in the end we found only 13 before funds ran out. There is something very wrong here. With an estimated 3 million or more claimed abductees in the US we could not find 15. Our requirements were high, but not that high. Nearly any medical study can find 1%; we could not find 1/1000%. This may say more than [the results of] our study."[133]

In September 2004 Deuley and Rodeghier informed me that progress in the analysis of the AMP data had been slow due to limited funding, their difficulty finding an independent statistical analyst, and the limitations of time, resources, and circumstances of the principal technical party involved in the study.[134]

One way or another, sooner or later, this sort of "crime scene" approach is bound to enrich our understanding of the alien abduction phenomenon.

An "Alien" Claw

Of course, the crime scene approach can be time consuming, costly, and ultimately disappointing. The case of Gary Lowrey, a safety officer for the Bakersfield Fire Department in California, whose story came to light in 2000, is a good example. The Lowrey family claimed that they were in the midst of an intense series of visitations by purported extraterrestrials, some reportedly with clawlike appendages. This complex UFO saga yielded not only a mysterious "footprint" but a number of bizarre "alien" images via a CCD camera. The "aliens" appear in a few frames, or are only visible when the camera is stop framed, revealing image fragments of busts or heads of strange insectlike "gargoyles or gremlins."[135] Then in September 2000, Gary Lowrey found what appeared to be definitive proof of his claims, an object resembling a claw in a bedroom that was the scene of much of the intense anomalous activity. Thus, when this "claw" was found, he assumed it was related to the strange creatures that had been observed in the bedroom.

This "claw" was of interest to the biochemists working with me as part of APEG. But because the case had emerged from the hotbed of Californian ufology and had already been caught up in considerable controversy, I tried to encourage some independent enquiries to establish the bona fides of the case. Dr. Colm Kelleher, a biochemist and administrator of the National Institute for Discovery Science, agreed to look into the case. Although he could only undertake a limited enquiry, Kelleher, after consultation with Dr. Roger Leir, who had already made extensive enquiries into the case, felt the "claw" sample could be a good candidate for DNA study. NIDS supported the work and eventually the APEG biochemists received a part of the claw to examine. The complexity and difficulty of the DNA on this case is summarized in a report entitled "A Cautionary Tale: DNA Analysis of Alleged Extraterrestrial Biological Material—Anatomy of a Molecular Forensic Investigation," compiled by Dr. Kelleher on behalf of the three cooperating research groups: NIDS, A&S Research (Dr. Roger Leir's organization), and my research group, APEG.

"The investigation of this case went far beyond the 'business-as-usual' analyses usually afforded anomaly cases," wrote Kelleher. "The project evolved into a major molecular biology research project in its own right." In fact, "six separate rounds of subsequent DNA analysis, using different and sometimes very novel approaches, were carried out to bring this case to a conclusion. It was necessary to invent a new polymerase chain reaction using novel primers to the most conserved DNA sequences on Earth in order to finally resolve this case." Initially the multiple rounds of DNA analysis of the biological sample appeared to corroborate its reported anomalous (extraterrestrial) origins. But the "claw" turned out to have far less exotic origins than aliens.

Finally, wrote Kelleher, "painstaking DNA analyses and the use of bioinformatics methodology over a 12 month period by highly qualified teams of experts in three countries was necessary to establish that the biological specimen found in the house was a mundane terrestrial mollusk. Mollusks, particularly snails and slugs, secrete a thick mucus that contains multiple inhibitors of many of the common enzymes that are fundamental to molecular biology and DNA analysis, including polymerase chain reaction enzymes and those used in standard molecular cloning. Further, there is a relative paucity of mollusk DNA sequences, particularly from mollusks found in California, in global DNA databanks. These two factors conspired to lead this investigation down a false path for about a year. Ultimately however rigorous DNA analysis using a novel set of oligonucleotide primers for the polymerase chain reaction solved the puzzle.... Independent confirmation that the sample was a dried mollusk was obtained by an expert from the Los Angeles County Natural History Museum."[136] The anomalies originally reported by the family remain unexplained, however.[137]

This lengthy and very expensive investigation provides a good lesson for investigators. The incident and the analyses that followed strongly suggest the need for rigorously establishing the strength of the link between reported events and any artifacts found and thought to be related to them. This step is vital because potentially so much hinges on it.

It is also important to challenge the initial identifications of artifacts. For example, in the case of Peter Khoury, one of our first priorities was to establish whether the found hair sample was even hair in the first place. Looks can be deceiving, but a close-up microscopic viewing quickly established the characteristic morphology of hair. In the case of the "claw" sample, it seems that "claw" identification was never seriously challenged early on in the investigation, and by the time DNA work was contemplated, it seemed almost a given. Ultimately persistence paid off, but in hindsight a lot of extra work and expense could have been avoided. A comprehensive and more thorough point of discovery analysis at the beginning of any case investigation would assist in determining whether the evidence merits a detailed and costly study.

The case once again underscored the value of the scientific approach, however. One of Carl Sagan's last books, *The Demon-Haunted World*, was subtitled *Science as a Candle in the Dark*. It was Sagan's intention to equate the "dark" with pseudoscience. But I feel the phrase should have a broader reference. I think that science should be used to examine any unknown, and specifically that science should indeed light the way in the study of alien abductions.

CHAPTER TWELVE

The Light Fantastic

A S WE HAVE SEEN, ABDUCTION CASES CAN SOMETIMES PROVIDE the sort of physical evidence found in more traditional UFO cases, like the ground markings that may be left behind in UFO landings. Such evidence adds to the argument that abductions occur in our physical reality. I have undertaken extensive research into the topic of UFO "physical trace" cases and have argued that the evidence supports a physical dimension to the UFO phenomenon.[138]

I have been particularly interested in what are called "solid light" cases, in which UFOs appear to use light in a coherent and controlled way. In some cases, the light beams from UFOs take on a seemingly solid appearance, turning corners, being used to lift things, or sometimes functioning as a transportation "tunnel" from one place to another.[139] I feel there very well may be scientific pay dirt in such cases. Many of these "solid light" cases appear to have biological effects on witnesses and abductees. Some are benign, some are not so pleasant, and others are remarkably beneficial, having what appear to be healing effects. In the extraordinary UFO encounter of Dr. X in France during 1968, for example, the craft projected beams of light that may have been responsible for the remarkable "rapid healing" of an injury the witness had suffered years before. And after a seventy-three-year-old Argentine named Ventura Maceiras was exposed to a UFO beam of light in 1972, he experienced a remarkable array of physical effects, including enhanced IQ and new teeth! [140] If such effects are due to these light beams, then forensic evaluations of their effects at the biological level, via DNA testing and perhaps other biomedical analyses, may reveal striking information about the processes involved and the apparent technologies deployed. Knowledge of such a mechanism would be an extraordinary biochemical and biomedical breakthrough.

Only in recent years have we begun to be able to manipulate light in ways that seem routine in UFO and alien abduction accounts. The ability to bend and stop light in extraordinary ways seems almost standard fare in UFO technology. As recent news headlines reveal—"Researchers say they have slowed light to a dead stop, stored it, and then released it as if it were an ordinary material particle"—for us, it's cutting-edge science. [141] Budd Hopkins and Carol Rainey in their book *Sight Unseen* provide a sort of de facto "reality check," when they highlight the various scientific breakthroughs and discoveries which resonate or reflect the bizarre elements that have emerged in abduction lore over the previous few decades. In essence, the body of abduction lore as perceived by Budd Hopkins, David Jacobs, and others may gain credibility given that some of the startling advances of our own sciences seem to eerily approximate some of the unlikely aspects of the alien abduction mythos.

In the following case, the element of "solid light" is prominent. It also contains elements of discontinuity in the natural flow of events and the perception of time, along with a description of people being caught up in the strange "solid light" beam. All of these and other aspects are common in abduction cases.

More Than a Dream?

I was told this story by a man who was troubled by a bizarre experience he had on the south coast of New South Wales, Australia, at the beachside town of Kiama, in the early 1970s. I have spoken to the witness, who I will call Graham, on a number of occasions and met him in person for an extended interview. I found him to be a compelling witness who is grappling with the ontological issues that striking episodes often force us to confront. I have quoted from his prepared statement:

Awoken by a light coming into the room, I was too drowsy to do anything about it, I wanted to sleep. . . . I thought it may be an intruder, so I forced myself awake, to step over the baby and my two year old daughter sleeping on the floor beside me. When I got to the window I could see nothing unusual outside. . . . I laid down again and fell quickly asleep. Again the light came into the room. This time I jumped up quickly, wide-awake again, there was nothing unusual outside. Suddenly I saw a light beam white in color with a blue fluorescent tinge evaporating from it. Because of the luminescence of the light I was able to make out the shape of a flying craft from which the beam projected at an angle to the ground of about 75°.

The beam was about thirty feet long and about two feet six inches in diameter, given the craft was between the headland I was on and the next headland. Suddenly the beam, still only thirty feet long fell, like a perfect cylinder of solid light. It did not fall in the direction of gravity, it continued along the path of its own axis. The cylinder of solid light hit a caravan. Upon impact the light behaved like water, pouring over the caravan, over its roof, over its walls, over every nook and cranny of the van, like fluorescent paint from an electro, airless spray gun. The caravan illuminated completely for about three sec-onds then the light faded away. My attention was on the light. I could not see the craft anymore.

I rubbed my eyes and looked for the craft. It appeared slightly to the left of its original position with another beam of light, descending from it at a very slow speed; say about only three feet per second. When the beam reached a given length, longer than the first time, it began falling as before. This time it hit an amenities block and the light covered its surfaces completely illuminating it in the same way as the caravan. Again the light faded away.

From the same location, the craft let another beam go at an angle of about 45° to the ground level line. The beam was much longer than before. It reached the beach and illuminated approximately an area of sand forty feet at its widest. Inside the lighted area were two men standing motionless looking up at the craft. A young woman jumped up from sitting near a small beach fire and ran to stand with the two men. A second young woman was running backwards trying to brush the light off her arms and body. Then she too stood separate to the other three and also stared up at the craft. The light suddenly went out and I looked for the people. Has it taken the people I thought? Where, as I was marveling at the craft and light before, I now became angry, thinking it has terrorized that woman. It was not a good thing as I first thought. Now I could see the fire dimly glowing. I looked this way and that to see if any of the people walked in front of the fire, to prove they were still there. I fell asleep on my feet. When I awoke I was standing on the other side of the window, one hand on the window.

I looked outside the window only feet away. The craft hovered over the street in front of the house. It maneuvered very close to the window. I was impressed that it looked like a spaceship. . . . The metallic material it was made of appeared as though it was unpolished Zinc alloy. . . . It was as if it was made from one piece of metal about forty feet wide and ten feet high, which began to spin in one direction, then it stopped and spun for a shorter time in the opposite direction. Then it stopped spinning, hovering in a steady position above the skyline. There were no thoughts it belonged to the western world. . . . I blacked out.

When I came to, the craft was still opposite my window. I thought why was I meant to see that it had no welds or seams, it seemed to want to show me that. I looked at a window shape about six feet wide and two feet six inches high with curved corners. The metallic window shield suddenly disappeared and I could see inside the craft. I saw no fittings. It had flat vertical off white walls. I felt very peaceful. A man walked into the room of the craft and stood in front of the window. As he walked in he was looking at a flat object he was holding in his hands, like a clipboard but thicker. He began to move his arms as though he was working on something at bench height below the window. Totally absorbed, he worked away. I felt completely safe. Another man then entered the room looking at the other man and what he was doing. He stood also facing me looking at the bench and pointing like without words he was helping the other fix something.

They had bright silver one-piece suits like thin wetsuits on, with no badges or markings. They carried no weapons or tools. . . . They didn't [seem to] know I was watching them. With that thought the last one to enter the room smiled at the other, then they both smiled directly at me. I had physical fright, my hair stood on end literally and I knew what it meant to be really scared. I dropped to the floor and said, "Everybody keep down. Stay out of the light." I knew that in the light they could control my thinking to feel and think peacefully. Suddenly great noise and severe vibration of the house took place. The laundry light went dim, the fridge began jumping about and there was great noise above the roof. The washing machine was bumping about also. I said, "Quickly get under the doorways, the house is going to fall." It was like the craft overhead sucked the electricity out of the house, then took off.

Bill shouted out, "Shit, what was that? It took the bloody roof off." I said, "It was a UFO." . . . Bill said, "Yes I saw it as it took the roof off." Bill was trying to comprehend how come the roof was still there. Gordon —— living behind came out to his back door and said very explicitly, "What the . . . was that? I thought it took the roof off." He too was greatly concerned with checking out his roof, reassuring himself it was still there. The lady next door on the seaward side opened a window and said, "Where did it crash? Do we have to get out?" She became very angry saying again, "Do we have to get out?" No you're safe. It's gone. Relieved, she said, "I am alone with the children tonight, that bloody pilot should be shot for that." It was unusual to hear people who I had never heard swear before, swearing.

Bill and Gordon were saying it was a UFO. Suddenly, ——, Bill's wife began to try and quiet everyone down and get us all inside. We decided that I should phone the Nowra base. I spoke to the duty officer. He said he was the only one on duty. He asked me if I saw any orange lights. I said, "Yes." He then quickly said it was a weather balloon you saw, it was let go at such and such a time from Jambaroo, it didn't inflate properly and other people reported seeing it as an orange light over Kiama. In my mind I thought, he knows what it was, it must be secret. I've done my duty reporting it, so that was that.

The next day [our wives] said two men in dark suits with ID tags came to the door asking did any one see anything unusual last night. Frightened by the men, they said no and the men went on. [Our wives] warned me not to speak about it, they were very frightened that something would happen to me if I spoke up and also it would make us a laughing stock in the community. The plan was we would forget it, not talk about it, even to one another. So it would be distanced from our lives.

Bill was reading the paper some days later and said an expensive Navy helicopter flew from Jambaroo over Kiama. It lost its electronics and crashed forty kilometers out to sea off Kiama. The navy was reported to be trying to recover it to find out what happened. The crew was rescued. I said, "Yeah, I know about losing power, the same thing happened to the helicopter as what happened to the fridge and the laundry light. The UFO took its electricity." Nothing was said further. We ignored the event. What I saw holds future understanding for me, if it was a dream I believe. Possibly it was an active imagination, a dream and actual occurrences combined.

Graham has since pondered the nature of this strange event. He is troubled by it and now feels more comfortable referring to it as a dream. The fragmentary nature of the events and the strange element of the experience beckon this interpretation. But there are also many startling elements in the experience that reflect the paradoxical reality of the UFO phenomenon.

When a "solid light" case involves a little physical evidence, a forensic approach can draw some metaphorical blood, as this next case clearly illustrates.

The Abducted Gundiah Crime Scene

In early October 2001 an alien abduction story involving a "solid light" made headlines throughout Australia and began attracting interest from around the world. The investigation of this controversial case followed a classic forensic "crime scene" approach involving three different police jurisdictions. The incident had occurred on Thursday, October 4, 2001, at a Gundiah property near Tiaro, in southern Queensland, and concluded in the early hours of Friday, October 5, 2001. The next day Diane Harrison, the well-respected director of the

Australian UFO Research Network, called and filled me in on the rapidly unfolding events. Based near Brisbane, in southern Queensland, she had been talking to the key participants. Diane asked if I wanted to get involved in the investigation, and I quickly agreed. We knew it wouldn't take long for the "UFO bandwagon" to uncritically embrace the Gundiah case as a cause célèbre. We were determined to try to unravel the events—whether fact or fiction.

The three people involved in the alleged events behind this extraordinary tale of abduction, teleportation, time distortion, aliens, "solid light," and physical traces were Keith Rylance, thirty-nine, his wife Amy, twenty-two, and their business partner, Petra Heller, thirty-five. Their Gundiah property was apparently being developed as the Whispering Winds winery, as well as a potential venue for motocross and other activities. The startling account that emerged from our interviews with them was fleshed out with primary source material such as TV interviews and a witness statement provided to police. This is the story they told.

Thursday, October 4, 2001

Keith Rylance told us that he had gone to sleep in the bedroom of his mobile home at about 9:30 p.m. Petra had retired to her bedroom in the annex of the mobile home. Amy stayed on a couch watching TV in the annex room they used as a lounge. These locations were in close proximity, separated by a window and wall respectively from the lounge. Petra's room had a door leading to the lounge, which was left ajar. The door to the bedroom, where Keith was, was apparently open to the lounge. Amy apparently fell asleep on the couch.

On this stormy night, at around 11:15 p.m., Petra was reportedly woken up. When she entered the adjacent lounge, she was confronted by an extraordinary sight, which allegedly quickly overwhelmed her. A bizarre rectangular beam of light was being projected through the open window of the lounge. This light beam appeared to be truncated at the end. Inside the beam, Petra claims to have seen Amy in a prone position, being carried out headfirst through the window. Underneath her, but still within the beam, were items that had been on the coffee table adjacent to the couch where Amy had been sleeping. Before fainting in shock, Petra noticed that the beam was coming from a disc-shaped UFO hovering just above the ground a short distance away, near a tree at the rear of a clear section of the property, immediately behind the mobile home annex.

A short while later Petra regained consciousness and began screaming. The commotion woke Keith, who stepped into the lounge and was confronted by a highly agitated Petra. The items that were originally on the coffee table were now on the floor in front of the window. He told us that he then found that the window screen was torn both vertically and along the bottom of the window frame. Unable to get any sense out of Petra, Keith rushed outside, trying to locate Amy. She was nowhere to be found. Keith then managed to get Petra to tell him what had happened. He initially didn't believe her. After a second search for his wife, Keith decided to call the police.

Keith called the nearby Tiaro police around 11:40 p.m. reporting that his wife had been abducted and asked for help. Due to the short staff at that time of night, it took about an hour and a half after the initial call for Senior Constable Robert Maragna from Tiaro and an officer from Maryborough to arrive at the site of the incident.

Friday, October 5, 2001

The police officers initially suspected foul play, perhaps even murder, until the bizarre circumstances of the alleged events came into focus. The two people, Keith and Petra, claimed that their companion, Amy Rylance, had been abducted by a "spaceship"! As the officers struggled to keep an open mind, they were joined by the officer in charge of the Tiaro police, Sergeant John Bosnjak, who had been asleep when the police called him to assist in the investigation.

The three officers then investigated the site. They had found Keith Rylance and Petra Heller in an agitated state. There was no sign of Amy Rylance. They examined the torn screen. The right side of a flowering bush, commonly known as "yesterday, today, tomorrow," located to the left side of the window with the torn screen, looked as if it had been affected by heat. The police took samples for possible later testing.

While the police were at the property, the phone rang and was picked up by Keith. A woman was calling to say that she had taken a somewhat distressed and apparently dehydrated young woman from a BP gas station on the northern outskirts of the central Queensland city of Mackay, some 790 kilometers by road to the north of the Gundiah-Tiaro area. The young woman turned out to be Amy Rylance, and the female caller explained that Amy was apparently all right, and was at the Mackay hospital, where a doctor had examined her. Keith handed the phone to Senior Constable Maragna.

Given these extraordinary circumstances, Mackay police were called in, making a total of three police stations involved in the investigation—Tiaro, Maryborough, and Mackay.

Keith Rylance gave the police the keys to the property,

packed up a substantial amount of their business and personal gear into a van, and set off for Mackay with Petra.

Meanwhile, at Mackay, Constable J. A. Hansen, of Mackay police, interviewed Amy Rylance, around 2:30 Friday morning. Amy completed a written statement, notarized with a Justice Act, attesting that the statement was true to the best of her knowledge and belief, and that if admitted as evidence, she would be liable to prosecution if she had indicated anything in it that she knew was false.

This statement indicated that her last recollection was of lying on the couch at the Gundiah property. She had no awareness of the events that Petra described, but claimed her next awareness was of waking up lying on a bench in a strange rectangular room. Illumination came from the walls and the ceiling. She was alone. She indicated she called out and heard what seemed to be a male voice, asking her to be calm and that everything would be all right and that she would not be harmed. Soon an opening appeared in the wall and a slender "guy" about six feet tall and covered in a full bodysuit walked into the room. He appeared to have a black mask on his face, with a hole for his eyes, nose, and mouth. He repeated his calming assurances. Amy felt she had been there awhile. The "guy" told her they were returning her to a place not far from where they took her, because the lights were wrong at the property and it wasn't safe.

The next thing she recollects is waking up on the ground with trees around her. She felt disoriented. She could smell the ocean. She was not sure how long she tumbled through bushland—it seemed to be a long time—but she felt she wasn't making much progress. She then came out onto a road that looked like a highway and saw a light from a gas station. She walked into the station, where the staff, seeing her state, offered assistance. She accepted some water, as she felt somewhat dehydrated. Initially she was unable to answer questions, and didn't know where she was. She was also asked if she had been drinking or was on drugs, to which she said no. Amy indicated she felt tired, sore, drained, and lethargic. She asked a woman at the service station to take her to the hospital, as she didn't know where else to go. The woman and her friend took Amy to the hospital.

Later Amy spoke with two police officers and also spoke with her husband Keith from the hospital. She then went to Mackay police station where she gave her statement. Amy also indicated that this sort of thing had never happened to her before, but when she was in the fifth year of school she had seen a large UFO surrounded by smaller objects.

The police arranged to put Amy in a motel pending the arrival of her husband. He and Petra arrived during the day and spent considerable time with Amy discussing what happened. They reportedly took extensive notes as well as photographs of a triangular arrangement of marks on her inner right thigh, marks on each heel, and the growing out of her hair, which she had dyed earlier in the week. Her hair had apparently started to show its former color, suggesting that considerable time had passed, certainly more than a few hours. Her body hair had allegedly also become somewhat more pronounced than would otherwise be apparent for the short time involved.

Keith Rylance said he then began to learn about the UFO subject on the internet at a café and bought a copy of *Australasian Ufologist* magazine in Mackay. He then contacted the Australian UFO Research Network number mentioned in the magazine. Diane Harrison took the call and for the next hour or so listened to the story that Keith and Amy told. Petra was apparently sleeping at the time.

I put a call through to Keith at the motel, securing permission to record the conversation. Once again Petra was not available to talk about her part in the alleged events. Keith Rylance went into considerable detail about the events, referring often to the notes they had apparently compiled during the day. The details described covered the events Petra had witnessed, what Keith had experienced, and what Amy told them had happened to her during her experience. Finally, I spoke with Amy, mainly about the events before and after the claimed onboard experiences, because Keith had already gone into considerable detail about the latter.

Keith Rylance seemed to want to control how both media and investigators would get involved. His desire to contact the media promptly led both Diane and me to suggest that he should think very carefully about the possible ramifications of doing so. Keith seemed to feel that it was important to get the story out, as it would come out anyway and this way he could control the way it did. He was also trying to restrict the way the investigators could or should look into their experience. He claimed they didn't need to prove the experience. While he didn't directly witness the experiences, he now believed both Amy and Petra. I explained to Keith that we wanted to look into the situation very thoroughly. I also explained how certain basic sampling procedures we could do with Amy's marks would be important in verifying their story.

Since the story seemed destined to be a big one, Diane and I decided to undertake a detailed on-site investigation. Keith Rylance had told us that the three of them would wait for us to come to Mackay, as they were in no apparent hurry to return to Gundiah. They gave us permission to visit the property in Gundiah on the way.

Tuesday, October 9, 2001

Diane and I traveled to Gundiah, arriving at the Whispering Winds winery property, just after 10 p.m. Because of the late hour, we obtained permission from the witnesses to stay there overnight. The next day we continued our investigation. Keith had arranged for a neighbor to regularly check on the two pets left behind, namely a parrot and a kelpie dog. He indicated to us that it would be okay to let the dog off for a run. When we let the dog off, we observed its behavior. At one point it jumped up on the window with the damaged screen. This gave some support to the possibility that at least some, if not all, of the damage, could have been caused by the dog, and closer inspection confirmed this. We also inspected the damaged plant and found a possible prosaic cause for it as well—simple heat stress from hot sunlight. A healthy flowering bush of the same species at Mt. Bassett cemetery in Mackay revealed similar damage. We also spoke to the police about the case.

Wednesday, October 10, 2001

In Mackay, we focused our attention on the area where Amy Rylance returned and tried to reconstruct the circumstances of Amy's return. We spoke with the BP gas service station staff that turned over surveillance videotape that might contain Amy's visit. The more we investigated the case, the more questions we had about it. Keith Rylance indicated to us he would be available for questions when we got to Mackay, but it became clear to us early on the first day of our investigations there that this was probably not going to be the case. When we contacted the motel where they had been the night before, their third motel in Mackay, we learned they had al-ready checked out that morning.

Thursday, October 11, 2001

We left messages on Keith's mobile phone but didn't hear from him until early in the afternoon of our second day, as we were leaving Mackay. He apologized for not being available, but indicated they had relocated to an unspecified location after having fled the area following a kind of "men in black" experience. Keith reported being pursued by a high-powered, dark brown four-wheel truck, which they managed to evade. We never managed to sample Amy's marks. Keith fell back on his mantra that this sort of evidence was not necessary to prove their case.

Monday, October 15, 2001

We heard from Keith Rylance again, but we were largely un-successful in getting him to answer some of the questions and issues raised in our investigation. We made him aware that Tiaro police would like to hear from him. Our prelimi-nary investigation report, which was completed by October 14, was circulated on the Internet and published in the *Aus-tralasian Ufologist*.[142]

Thursday, October 18, 2001

I spoke with Keith Rylance again. He recanted his men-in-black experience, which he had told me had caused them to flee Mackay. Keith also told me he had spoken with the Marybor-ough police, but I responded that all investigation had been

handed back to Tiaro police, the original investigating station.
I suggested again that they would like to speak with him.

Friday, October 19, 2001

Keith Rylance called. He was clearly angry about the preliminary report Diane and I had written. He said that we had got
it wrong and cited errors, namely, we had given his age as 40
(he was 39) and Petra's age as 39 (she was 35). Petra said, "I
forgive you." I said we had gleaned that information from
media reports, but that the actual case data was accurately reported, given our limited access. Keith then mellowed a bit
and agreed to answer my questions by e-mail so we could
move on and start talking about the events, that is, the craft,
the alien, and so on. We had hoped to speak with Amy and
Petra separately and privately, in much more detail, but Keith
was omnipresent as the primary source of information.

November 2001

The mass-circulation women's magazine *New Idea* came out
with a four-page story on the affair, which revealed many of
the details Keith and Amy Rylance had conveyed to us. In the
article Amy was quoted as saying, "I didn't do this to grab
media attention. I know it sounds far-fetched, but I know it's
real and that's all that matters."

The author of the article wrote: "[We] were in contact with
Keith and Amy and even slept at their Gundiah house. Bill,
considered one of the best researchers in his field, spoke to
Keith about his desire to take DNA and blood samples from
Amy. He argues that obtaining biological evidence is crucial to
support her claims the triangular puncture marks on her thigh

are evidence of alien experimentation. However, Amy, Keith and Petra then fled Gundiah, and the UFO investigators haven't been able to complete their research." The writer then quoted Diane as describing the case as "extraordinary and controversial," and saying, "There are too many unanswered questions to draw any real conclusions. We want to keep an open, objective mind before we conclude one way or the other."[143]

December 2001

Keith Rylance's contact details, mobile phone and e-mail, were no longer functional. Diane and I were resigned to the fact that we would probably never get any responses or answers to the numerous issues and questions we had. We were now highly dubious of the case.

October 2002

On the anniversary of the alleged event, the *Fraser Coast Chronicle* newspaper asked, "Why did they do it?" The story argued that the affair had been a hoax and highlighted some of the numerous forensic issues we and the police had found back in October 2001, and had unsuccessfully been seeking answers to from the trio.

The Tiaro police were not so circumspect: "The police file remains open and Sergeant Robert Maragna [of Tiaro police] would love to talk to Amy, Keith and Petra and give them the bill for the hours of police time that went into investigating the alien abduction. 'There were too many inconsistencies in their story for it to be true,' said Sergeant Robert Maragna. During a search of the property police found black hair dye, paper towels and the burnt out remains of two flood lights and electrical wiring in an incinerator about 20 meters from the annex. 'The most damning evidence are the phone records,' said Sergeant Maragna." Calls came from a motel in Rockhampton, which is between Gundiah and Mackay, to the Gundiah home the day before. The police scenario had Amy already on her way to Mackay via the Rockhampton motel, with the black dye being used by Petra later to play the blond Amy. The newspaper also quoted the Tiaro mayor John Horrex, however, who still believed the trio's story of an alien abduction.[144]

March 2003

We were contacted by an overseas source who suggested the saga had more to do with some misguided secret Scientology "mission" to find a buried spaceship, as described in the *Mission Earth* science fiction series by Scientology's founder, L. Ron Hubbard. The scenario seemed unbelievable, but the source claimed intimate knowledge of the incident. However, when we were at the Tiaro police station back in October 2001, we were shown a notebook, apparently owned by Petra, which the police had removed from the Gundiah property. The police were suspicious, as it seemed to have some strange content they thought might be connected to the weird story under investigation. I recognized some of its content as being related to Scientology, the words "Thetan" and "clearing," for example, and then thought that it probably wasn't relevant to our inquiries.[145] But given this new information, maybe it was. Recalling that the Gundiah property was called Whispering Winds, Diane mentioned that Hubbard, lived out his last years on a remote property in the United States—"It is a quiet place, a perfect place to hide."[146] It's name? Whispering Winds!

CHAPTER THIRTEEN

Strange Speculations

T HE STRIKING RESULTS THAT EMERGED FROM THE DNA PRO-file of the biological sample recovered during Peter Khoury's 1992 experience seem to confirm that he had a bizarre encounter with strange beings with a very rare genetic fingerprint. Though there is a limit to how much information a single sample can provide, the nature of these genetic findings led us naturally to speculate about their possible connections with three ancient cultures: Chinese, Basque, and Gaelic.

Dragon Seed—the Chinese Connection

The initial mitochondrial DNA findings from the shaft of the Khoury hair sample confirmed that the hair came from someone who was biologically close to normal human genetics, but of an unusual racial type—a rare Chinese Mongoloid type—one of the rarest human lineages known and one that lies further from the human mainstream than any other except for African pygmies and aboriginals. But this result emerged from a *blond* hair sample, rather than a *black* one, as would be expected from the Asian type mitochondrial DNA. How do we reconcile this with the description of the tall blond female that Peter Khoury encountered? Hair dying and albinism were ruled out as possible explanations.

The DNA data led us first to China in a search for clues to this dilemma.

Western China attracted our attention, particularly Xinjiang province,[147] where a group of remarkably well preserved mummies emerged from the shifting sands of the forbidding Taklimakan desert in 1994. Even more impressive than their state of preservation was their appearance. The mummies did not resemble the Chinese or ethnic minorities that inhabited the area. They were far stranger still and very much out of place, for they are remarkably western or Caucasian in appearance, with fair hair of various shades: blond, dark blond, brown, and even straw orange. These people, who lived in western China a few thousand years ago, are generally much taller than the current inhabitants, and the rich fabric of their clothing startled many researchers.

I am not about to suggest that the Taklimakan mummies are in some way related to the Nordic-type beings often reported in alien abduction stories. Such a possibility can only be considered once a complete DNA profile of the mummies is available and the biological evidence from a significant number of abduction cases provides a consistent DNA profile. However, the complexity of ethnic movements and interplays in the Xinjiang region—part of the Silk Road trade route—means the mummies are but a tantalizing and elusive fragment within a burgeoning genetic jigsaw.

Victor H. Mair, the man largely responsible for the West's current reexamination of the Taklimakan mummies, is the coauthor, with J. P. Mallory, of the best book on the subject, *The Tarim Mummies: Ancient China and the Mystery of the Earliest Peoples from the West*. On the tentative unraveling of the mummy's DNA fingerprint, the authors wrote:

"It was his drive to know the results of DNA analysis on the mummies that first prompted Victor's involvement in the prehistory of Xinjiang. As part of the initial project, Paolo Francalacci from the University of Sassari obtained tissue samples from 11 mummies from Qizilchoqa and from mummies housed in the museums at Urumchi and Korla, as well as at the Institute of Archaeology in Urumchi. Although he sampled 11 mummies, the Chinese authorities permitted Paolo to carry off samples from only two of them, and the DNA in one of these was too damaged for analysis. At present the genetic history of the mummies rests with but a single individual.

"Paolo discovered that the DNA of one of the mummies belonged to what is known as Haplogroup H, one of the nine subtypes of mitochondrial lineages that are largely associated with European populations. Haplogroup H[148] is the commonest marker of European populations and occurs in about 40 percent of Europeans (but also in about 15 percent of people from the Near East), while haplogroups A through G are more typically 'Asian.'[149] As it may be found among people of such diverse ancestry as Swedes, Finns, Tuscans, Corsicans and Sardinians, it cannot so far relate the mummies yet to any specific European subgroup; it merely emphasizes once again that the Xinjiang mummies find their closest genetic relations among Europeans. As for Victor's original quest: the relationship between the Tarim mummies and the Tyrolean 'Iceman,' the latest results on the latter show him to belong to Haplogroup K, a widespread European haplotype which occurs in about 10 percent of the population of Europe. Paolo is convinced that if he were able to recover more tissue samples from other Tarim Basin mummies, they would probably reveal other European and Asian haplogroups." [150]

It will be some time before the full genetic jigsaw of the Tarim/Taklimakan mummies is fully resolved. Until then, these out-of-place, tall, fair-haired people cast an evocative spell over the mystery being studied here, a mystery, needless to say, that resonates powerfully with some rather seductive tales from China's rich past.

Lost with the Lights

Travis Walton's five lost days courtesy of a UFO abduction back in 1975, or Credo Mutwa's three missing days back in the late 1950s, are but winks in time compared to a legendary Chinese tale that prefigures a Rip Van Winkle–like trip into "cave heavens." Tu Kuang-t'ing (858–933 A.D.) in his "Report Concerning the Cave Heavens and Land of Happiness in Famous Mountains" lists ten "cave heavens" and thirty-six "small cave heavens" that were thought to be under Chinese mountains. Tu relates the mythical tale of a man who entered one of the "cave heavens":

"After walking ten miles, he suddenly found himself in a beautiful land 'with a clear blue sky, shining pinkish clouds, fragrant flowers, densely growing willows, towers the color of cinnabar, pavilions of red jade, and far flung palaces.' He was met by a group of lovely, seductive women, who brought him to a house of jasper and played him beautiful music while he drank 'a ruby-red drink and a jade-colored juice.' Just as he felt the urge to let himself be seduced, he remembered his family and returned to the passageway. Led by a strange light that danced before him, he walked back through the cave to the outer world; but when he reached his home village, he did not recognize anyone he saw, and when he arrived at his house, he met his own descendants of nine generations hence. They told him that one of their ancestors had disappeared into a cavern three hundred years before and had never been seen again." [151]

The county annals of Song-Zi Xian hold a similar and fascinating, although not so legendary account, from Hubei province:

"A certain Mr. Ju Tan, a farmer of Xi-Yan, arose early on the morning of May 8, 1880, and was walking along towards a copse on the hill behind his cottage, when he caught sight, in the undergrowth, of a strange glowing object, shining brightly in many colors. He stepped forward and attempted to seize hold of it, and suddenly felt himself swept up into the air, as though beyond the clouds, with a loud sound of wind in his ears, his mind stupefied and numbed, and quite unable to move his body. Then suddenly he was conscious of dropping from a great height, but onto a lofty mountain-peak.

"Feeling as though he was awakening from a dream, the farmer was terrified beyond measure. Some time later a woodcutter came along and questioned him as to who he was and where from, and Ju Tan replied 'I am from Song-Zi County in Hubei Province.' The woodcutter exclaimed: 'How did you get here! This is in Guizhou Province, more than a thousand li from your place.' [152]

"The woodcutter showed him the shortest path down the

mountain. Ju Tan arrived in due course at his home, and found that eighteen days had elapsed. After much enquiry, it was still impossible to know what had caused his experience. It was just an extraordinary mystery." [153]

These stories are echoed in modern accounts from China.

"One evening in the autumn of 1975," according to an unverified report published by the Beijing branch of the Chinese UFO Research Society, "two soldiers of a certain unit of the Chinese People's Liberation Army stationed in Jian-Shui County in the Province of Yun-Nan [Yunnan] encountered a huge saucer-shaped flying object that was circling around above their heads and emitting beams of soft orange-colored light. One of the men at once ran into the camp to give the alarm, while the other stayed there to watch it. A few minutes later, when the Camp Commandant with about a dozen armed men came running up to the Entrance to the Barracks, they found no trace of the solider who had remained behind. The Commandant at once ordered that a general search be made by all personnel, officers and enlisted men, but there was no sign of him. A few hours later, four soldiers, taking over sentry duty, suddenly heard the sound of someone moaning behind them, and, looking around, found that it was the missing man, who in miraculous fashion had reappeared. They at once perceived that his eyebrows, beard, and hair had grown extremely long. When he had fully recovered consciousness it was found that his memory was totally gone. His wristwatch showed that it had stopped long ago. It was not a watch of the type indicating the date, so it was not possible to establish how many 'days of a different time' [154] he had spent 'elsewhere.' His weapons and watch all were found to be slightly magnetized." [155]

During the Sydney Writers' Festival of May 2003, I had the

opportunity to meet Ma Jian, a literary dissident offspring of the Cultural Revolution's "lost generation." He had written a beautiful and disturbing book, *Red Dust: A Path Through China*. He told me of his strange encounter with a mysterious light near the Burmese border, again in Yunnan province, in about early May of 1986. In the middle of the night he had escaped from police custody in a small village, where he had been held under suspicion of being a dissident or spy.

"I left the village and climbed the hill, not daring to turn my torch [flashlight] on. Soon I could hear the river again. When my hand touched a concrete bridge I switched on my torch and ran.

"Half an hour later the batteries ran out and I was plunged into darkness again. Cold shivers ran down my spine. I was on a narrow path on a high mountain ridge. One false step and I would roll to my death. I could hear the wind rustle through treetops in the valley far below. I wished I had the eyes of a mouse and could see in the dark. I crouched down and began to crawl like a pig. . . . I groped like this for hours until I could go no further. At last I collapsed and sank my face into the ground. . . .

"At that very moment a light appeared in the darkness. It was neither a torch nor a candle, nor a glow-worm shaking in the breeze. It seemed to come from another realm. It rose from a stream and floated through the trees then stopped by some branches ten meters away and slowly dropped to my eye level. I shut my eyes and tried to compose myself. Suddenly I remembered a story my father told me as a child. One night, when my grandfather had lost his way in a field of sorghum, a ball of fire appeared before him and guided him back home.

"I opened my eyes and stood up. He is here. My grandfather who died in a communist jail before I was born has come to my rescue. I walked forward and the ball of fire followed me through the branches, guiding my way for twenty kilometers until the sky turned white."[156]

Ma Jian confirmed that this account was not a literary metaphor for his sense of alienation with contemporary China. It was a real experience with a ball of light about the size of a cantaloupe.[157]

The Harbin Abduction

In September 2003 *Time* magazine reflected on China's imminent first manned launch into space. "Forget China's astronauts. The country's most famous intergalactic traveler lives in the last house on his lane at the edge of a Siberian forest," *Time* reporter Matthew Forney enthused, tongue in cheek, I suspect.[158] The best-known alien abduction story in China is the case of Meng Zhao Guo,[159] a young tree farmer from Wuchang, near Harbin in Heilongjiang province. In June 1994 Zhao Guo and two other farm workers at Red Flag logging camp saw something unusual on nearby Mt. Phoenix—"a metallic glint. . . . Thinking a helicopter had crashed, he set out to scavenge for scrap."[160] The trio climbed the mountain and were able to see that it was "a strange, white round object with a tail like a scorpion."

Sun Shi Li participated in the Chinese UFO Research Organization's (CURO) investigation of the affair.[161] In 1997 he indicated to American researcher Antonio Huneeus, "[Zhao Guo] couldn't get closer because it was emitting a very strong noise that produced unbearable pain. So they changed direction to approach the object again but with no success. . . . On the following day he guided a group of co-workers to the site and, when he was about 1 kilometer away, he looked with binoculars and saw next to the device an ET with a raised arm emitting a beam which burned his forehead. . . . He fainted, falling to the ground." [162]

Time's belated take described what followed. Several nights after his injury, "Meng says he found himself floating above his bed. As his wife and daughter slept below, a three-m-tall, six-fingered alien with braided fur on her legs straddled his waist. After forty minutes of levitational copulation she departed through the wall, leaving Meng with a five-cm mark on his thigh. A month later, he says, he was transported through the wall into a spaceship. Meng asked to see the woman with the braided fur. Impossible, they said. But they gave him hope. 'In 60 years, on a distant planet,' they said, 'the son of a Chinese peasant will be born.' Meng asked if he would ever see this child. He would. The aliens did not say where." [163]

Huneeus indicates that the story becomes even weirder: "When Zhao Guo was taken by train to the hospital, he claimed seeing an invisible alien—a three meter tall, scary-looking female ET—which nobody else in the train could see. He even claimed having sexual intercourse with her, which CURO investigators have a hard time believing. That same night, Zhao Guo took a photo outside his house and an unexplained white bar appeared in one corner when the film was developed." [164]

Sun Shi elaborated on the more exotic aspects of the case in a lecture he gave in Brisbane, in Australia, during 1997: [165]

After being hit by the ray he had a rare experience with an ET female. The female took him through a wall. She took him at night. He felt cold, so the ET gave him a cover, which was also a flying vehicle.

In the abdominal area she inserted two grains of medicine, and also in his leg.

[Zhao Guo] lost his memory of everything that happened. After this he recovered his memory.

The ET female was very tall, almost roof height. She showed him some images on a screen that was on the eve of a crash into Saturn. The ETs predicted this crash. . . . [166]

I suspect that the "crash" being referred to here is the crash of Comet Shoemaker-Levy 9 into Jupiter on July 16, 1994. As Zhao Guo had his claimed encounter during June 1994, it seems likely that this was what was being implied. But, in fact, astronomers had been predicting the collision of the comet with Jupiter for some time.

He was also shown images of Earth and of President Clinton and Chinese dignitaries. Their technology is very advanced. It can reflect anything on Earth.

[In the farmer's house] he had sexual relations with her. . . . The ET female maintained sexual encounters on a variety of occasions. [Zhao Guo] has sketched in detail the anatomical parts of the female. . . . He was told a baby of mixed blood [would be born]. Sixty years later she will return to show the child to the farmer.

After the event an orchid by the window . . . after several nights, this orchid gave off brilliant rays of light for several days.

During September 2003 Zhang Jingping, a Beijing-based UFO researcher, had psychologists and police technicians subject Zhao Guo to hypnosis and a lie detector test in Beijing. Zhang indicated the test results proved the abductee was telling the truth. He also claimed that doctors had indicated that Zhao Guo's scars "could not possibly have been caused by common injuries or surgery." Meng Zhao Guo, a humble farmer with only five years of schooling, also indicated that he had never heard of UFOs or ufologists until after his experiences had been reported. After more than a hundred interviews he now feels the affair has disrupted his life. Somewhat uneasily, he laments, "But ufologists still take great interest in [my] UFO encounter nine years on. They hope there will be a conclusion to the UFO phenomenon as soon as possible; only then will I feel released."[167]

Like so many of the worldwide cases of alien abduction, Meng Zhao Guo's story is essentially just that—a story, and a bizarre one at that. To really establish the nature of the reality behind it, further research is required, and that research, whenever possible, needs to be anchored in the forensic/scientific approach I have advocated throughout this book.

Basque Legacy—the Iberian Vinculum

Further DNA tests on the Khoury hair sample revealed a rare Basque/Gaelic type DNA in the root section. Following each of the localities suggested by the genetic evidence, I turned my attention to the world of the Basques.

The Basques are believed by many to be Europe's oldest people, but their origins are a mystery. Their language is thought to be unique, largely unrelated to any other language on Earth. They are few in number and inhabit a rugged and remote part of the European mainland between France and Spain.[168,169]

Mari, the neolithic goddess of Old Europe and the primary deity in Basque mythology, has many manifestations, including "as a tree that looks like a woman or a tree emitting flames," "a white cloud or rainbow, or a ball of fire in the air," a "sickle of fire, as which she appears crossing the sky," and "seen enveloped in fire, lying down horizontally, moving through the air."[170] Her preferred habitat is generally subterranean, as befits the Basque lands, which are wild and mountainous.[171]

So the Basques are a unique and unusual people with a mythic belief in a potent female "sky being," often perceived as a light form moving through the air. Modern abduction reports in the wider Iberian Peninsula are no more tangible than this myth, unfortunately.

Trolling under the Dolmen

Many Iberian abduction cases were made public by the late Iberian ufologist Antonio Ribera. One such case involves the bizarre claims of Xavier Clares Jerez, who claims not to have been abducted into a UFO but into a cave, which resonates well with the Basque goddess Mari's subterranean predilections.

On Sunday, July 21, 1985, Xavier Clares Jerez returned home after spending what he thought had been a relatively uneventful photographic day excursion to Vallgorguina. At home he found his wife and brother frantic: he had been gone for thirty-four hours instead of several hours, as he thought. His car was very dirty and was covered "with stains of some viscous, sticky substance." Though he thought he had traveled only fifty kilometers in his car, the odometer revealed three hundred kilometers. And to top it all off, he was sporting a bizarre haircut.

Totally confused by the turn of events, he turned to his cameras in the hope of unlocking the mystery. He had taken two cameras with him—a modern 35-millimeter camera loaded with color slide film and an old-fashioned box camera he had made himself. Though he didn't recollect taking any photos, his 35-millimeter camera indicated that he had taken several shots. When developed, the slide film revealed a series of "green demonic faces."

Under hypnosis by a Professor Francisco de Asis Rovatti Heredia, Xavier Clares Jerez told of driving out on a road that leads out to the Gentle Stone *(Pedra Gentil)*—a dolmen, or megalithic monument—at Vallgorguina in the Catalonia region of Spain. The name Vallgorguina is thought to be a corruption of the Basque words *Val Sorgina,* meaning "Valley of the Witches." He described seeing "a bit of mist ahead," which puzzled him. He then found himself dozing off on the ground. "A sort of liquid is falling on me. . . . It is sort of sticky . . . ugh . . . very nasty . . ." Confused, he said he wanted to sleep, but "I can't; this water that's dropping won't let me. It's certain it isn't water. . . . And everything's covered in clouds." He then took some photos of "the sky and this rain that is falling" with the box camera. He was protective of the camera, as "the spool I am carrying is important." He added, "I don't want them to take it off me . . . I can see them."

Xavier indicated under hypnosis that these beings were not like the small, benevolent ones he has encountered in his home. He doesn't like these beings, but they tell him, apparently telepathically, that they want him to go with them *to see how he is.* He described being reluctantly taken down a slope and into a cave. They tried to open his box camera, until one of them seemed to understand Xavier's concern. They then stuck something like a catheter into one of his arms and extracted some liquid.

Xavier described his abductors: "These are almost as tall as I am, just a bit less, but their faces are horrible. . . . They are deeply furrowed. You can't see any clothing on them. . . ." The one who put the "catheter" in him had hair on his head, but the others did not. They were "a very dirty gray" color.

Still under hypnosis, Xavier said the beings tried to use both of his cameras. He is told that they intend to make a copy of him. They even cut his hair, apparently because "they want to copy it." He apparently lost consciousness. Xavier then spoke about the strange company—his mirror-image copy—who drove back in the car with him. He was identical—even wearing the same clothing—except that his double had long hair, while Xavier's was now short, courtesy of a very bad alien barber. Xavier dropped off his "copy" upon arrival back in Barcelona.[172]

Xavier has his alien stigmata, like many others before and after him—"three triangular punctures" forming a triangle with approximately 3-centimeter sides on his left arm. He also made a detailed sketch of the three-pronged alien device that produced the marks. Xavier's box camera yielded photos showing "a black shape and two claw-like hands, black, shiny and scaly." Allegedly the original plates were very dark, forcing Xavier to make "a contra-type," or negative, apparently, followed by several generations of prints, each one successively cleaner and with less contrast. The end result shows a helmeted being but with little facial detail, plus some "hand/claw" shots.

Over the next few days there were a number of sightings of Xavier's "copy." His own aunt berated him, when he turned up for dinner: "What—you again! But you have just left, saying that you couldn't stay for dinner." Perhaps understandably, if this story is to be believed at all, Xavier became very preoccupied with the whereabouts of his "double." Intermittent sightings continued. With the approaching cold weather, a dejected-looking "double" wearing the same flowered Hawaiian shirt Xavier had worn in July was spotted at a bus stop on November 16, 1985. On January 12, 1986, just before Ribera submitted his report for publication, with Barcelona in winter's full grip, Xavier's "double" was seen again, still wearing that Hawaiian shirt![173]

As ridiculous as this story seems, there are some interesting resonances with other encounters around the world. Xavier Clares Jerez's bizarre tilt into the "alien underworld" began with the strange "rain," "mist," and "clouds." These are remarkably frequent in UFO events; in some cases, unusual precipitates are also reported.[174] Given my background in chemistry, I always hope for the chance of examining such material closely. I know of at least three such cases from Australia. In August 1971 in Gladstone and Rockhampton, Queensland, a "very thin film of odorless oil" was found on a car following a UFO encounter.[175] In 1974 two women from Canberra who experienced a missing-time encounter claimed that a "viscous sticky white material like spiders web . . . was all over the car door."[176] And a "thick white substance not unlike white paint" featured in the very strange car "road hazard" incident at Nemingha in 1976.[177] Unfortunately, I never had the opportunity to analyze these residues.

Under careful scrutiny, Xavier's bizarre tale just doesn't hold up, however. "For me," says Spanish UFO researcher Vicente-Juan Ballester Olmos, whose investigative rigor and skills are beyond reproach in my opinion, "it was a hoax/psychosis case, an unreliable testimony from a subject who was familiar as an insider to esoteric circles, an individual who was a witness of so-called bedroom visitations. In addition, sloppy hypnosis procedures were used. After Ribera, no one ever worried about the case and, to the best of my memory, no one came back to it again. Probably it was 'too much' even for gullible ufologists."[178]

However, as Vicente-Juan indicated, as far as we can tell no one analyzed the viscous material that allegedly fell on his car back in 1985, and there has been no independent analysis of his "alien" photos. So ultimately there is no credible evidence for the story at all.

Vicente-Juan himself reviewed six Iberian abduction experiences in a paper entitled "Alleged Experiences Inside UFOs: An Analysis of Abduction Reports." He found all six to be the result of hoax, delusion, or psychosis. "In none of the cases," states Vicente-Juan, "was extraordinary evidence presented to support an anomalous event or novel phenomenon."[179]

It is precisely this feature, "extraordinary evidence" supporting the anomalous events of Peter Khoury's and Kelly Cahill's experiences, that caused me to shift my opinion on abduction experiences. Now I think that abduction stories that yield extraordinary data that survive a forensic and scientific gauntlet deserve serious attention. The fact that some are emerging speaks loudly for the possible reality of alien abductions.

Tuatha Mythos—the Gaelic Celtic Nexus

DNA testing of the Khoury "alien hair" sample also yielded genetic links to the Gaelic peoples. The principal biochemical scientist who worked on the sample speculated about the DNA's possible Irish, English Gaelic, and Celtic origins. The most compelling speculations anchored around the mythic traditions of Ireland involving the mysterious Tuatha de Danann, or Tribe of Danu. The coming of these people to Ireland is recorded in the *Lebor Gabala Erren (The Book of the Taking of Ireland* or the *Book of Invasions),* compiled during the twelfth century A.D., but based on earlier sources. The Tuatha de Danann were apparently tall blond or red-haired strangers, "expert in the arts of pagan cunning," who supposedly interbred with the locals, while teaching them many useful skills. Their entrance to Ireland was dramatic:

"In this wise they came, in dark clouds from northern islands of the world. They landed on the mountains of Con-maicne Rein in Connachta, and they brought a darkness over the sun for three days and three nights. Gods were their men of arts, and non-gods their husbandmen."

It seems the Tuatha de Danann were advanced enough to arrive in western Ireland (near modern Connacht) by air. They were comprised of two social classes: as "gods" teaching medicine, smithing, communication, or druidry, and as "non-gods"—farmers or shepherds. Although no one knows for certain what the Tuatha looked like, descriptions, such as of their female war-leader, Eriu, indicate they were tall, attractive people with pale skin, high foreheads, long red hair, and large blue eyes. Other descriptions mention their blond, golden hair and blue eyes. These descriptions recalled not only the blond woman with red body hair in the 1957 abduction of Antonio Villas-Boas, but one of Peter Khoury's female visitors as well.

The records, which are steeped in myth and legend, indicate that the Tuatha were defeated and subsequently retreated. Apparently the "Ever-Lasting Ones," as they were known, managed to survive "as immortals in ancient barrows and cairns," according to James Mackillop, a noted expert on Irish studies.[180] "The world is then divided in two, the surface going to the ordinary, mortal Milesians and their progeny while the Tuatha live underground, the route to which is the sidh." The sidh are the "fairy mounds," and "by implication, the realm beyond the senses, the Otherworld or, in oral tradition, the fairy world."[181]

Some modern-day researchers of alien abduction lore find this duality of the Tuatha and the fairy world most intriguing. Both feature tall, often long-haired and blue eyed attractive peoples with striking abilities and resources (the "Tuatha" and the "Nordics") as well as diminutive beings often intent on entrapment and abduction (the "fairies" and the "grays"). I think this argues well for revisiting the validity of the links between the Tuatha legends, fairy folklore, and modern-day alien abduction lore.[182]

In one of the classic texts on fairy folklore, *The Fairy-Faith in Celtic Countries*, W. Y. Evans-Wentz provides an intriguing account from "a county Kerry Seer" of "a collective vision of Spiritual Beings," whose form and connection to light phenomena supports a direct parallel to modern-day UFO entity and alien abduction reports:

"Some few weeks before Christmas, 1910, at midnight on a very dark night, I and another young man (who like myself was then about twenty-three years of age) were on horseback on our way home from Limerick. When near Listowel, we noticed a light about half a mile ahead. At first it seemed to be no more than a light in some house; but as we came nearer to it and it was passing out of our direct line of vision we saw that it was moving up and down, to and fro, diminishing to a spark, then expanding into a yellow luminous flame. Before we came to Listowel we noticed two lights, about one hundred yards to our right, resembling the light seen first. Suddenly each of these lights expanded into the same sort of yellow luminous flame, about six feet high by four feet broad. In the midst of each flame we saw a radiant being having human form. Presently the lights moved toward one another and made contact, whereupon the two beings in them were seen to be walking side by side. The beings' bodies were formed of a pure dazzling radiance, white like the radiance of the sun, and much brighter than the yellow light or aura surrounding them. So dazzling was the radiance, like a halo, round their heads that we could not distinguish the countenances of the beings; we could only distinguish the general shape of their bodies; though their heads were very clearly outlined because this halo-like radiance, which was the brightest light about them, seemed to radiate from or rest upon the head of each being. As we traveled on, a house intervened between us and the lights, and we saw no more of them. It was the first time we had ever seen such phenomena, and in our hurry to get home we were not wise enough to stop and make further examination. But ever since that night I have frequently seen, both in Ireland and in England, similar lights with spiritual beings in them."[183]

Tuatha Visions in the Mystical Air

The Irish mystic, Æ, or George Russell, in his work *The Candle of Vision,* originally published in 1918, drew some remarkably cogent and meaningful "conclusions" about his own "visions in the air," which ostensibly took place about 1893.

"I was impelled to look upwards, and there above me was an airship glittering with light. It halted above the valley. . . .

"The apparition in the mystical air was so close that if I could have stretched out a hand from this world to that I could have clutched the aerial voyager as it swept by me."

These "visions" were thought to be of the Tuatha, revealing themselves and their "aerial ships," to Ireland's greatest mystic of the day. The points Russell makes about them are surprisingly relevant to modern accounts of "contact" and "abduction."

"I began to ask myself where in the three times or in what realm of space these ships were launched. Was it ages ago in some actual workshop in an extinct civilization, and were these but images in the eternal memory? Or were they launched by my own spirit from some magical arsenal of being, and, if so, with what intent? Or were they images of things yet to be in the world, begotten in that eternal mind where past, present and future coexist, and from which they stray into the imagination of scientist, engineer or poet to be out-realized in discovery, mechanism or song? I find it impossible to decide. Sometimes I even speculate on a world interpenetrating ours where another sun is glowing, and other stars are shining over other woods, mountains, rivers and another race of beings. And I know not why it should not be so. We are forced into such speculations when we become certain that no power in us of which we are conscious is concerned in the creation of such visionary forms. If these ships were launched so marvelously upon the visionary air by some transcendent artisan of the spirit they must have been built for some purpose and for what? I was not an engineer intent on aerial flight, but this is, I think, notable that at the moment of vision I seemed to myself to understand the mechanism of these air-ships, and I felt if I could have stepped out of this century into that visionary barque, I could have taken the wheel and steered it confidently on to its destiny. There is an interest in speculating about this impression of knowledge for it might indicate some complicity of the subconscious mind with the vision which startled the eye. That knowledge may have been poured on the one while seeing was granted to the other. If the vision was imagination, that is if the airship was launched from my own spirit, I must have been in council with the architect, perhaps in deep sleep. If I suppose it was imagination I am justified in trying by every means to reach with full consciousness to the arsenal where such wonders are wrought. I cannot be content to accept it as imagination and not try to meet the architect. As for these visions of airships and for many others I have been unable to place them even speculatively in any world or any century, and it must be so with the imaginations of many other people. But I think that when we begin speculation about these things it is the beginning of our wakening from the dream of life."[184]

Ireland, England, and Scotland are all rich in the Celtic tradition of the "Otherworld," where all manner of things intrude into our seemingly well-understood world. The landscape of ancient sites and monuments is often powerfully anchored in bizarre light and apparitional phenomena. Many of these locations have potent reputations today because they are seemingly haunted by UFO phenomena.[185]

The "Dark Angel"

Soon after completing the first phase of the DNA study of the Khoury sample, one of the APEG biochemists had a bizarre experience with a Celtic twist. He had rushed overseas to be with his father, who was reportedly near death in a rehabilitation clinic following a devastating stroke. The father was a high-ranking military man turned lawyer. He had no patience for nonsense, was profoundly nonspiritual, and had no interest whatsoever in New Age matters.

At three o'clock one morning, apparently at the point of death, he saw a woman appear beside his bed. It was dark and although he couldn't see her very well, she seemed to him to be of "average" appearance. But what followed was hardly average. When the woman took his hand and held it, he was overwhelmed by what seemed like a flood of "life energy" pouring into him. The woman then introduced herself: "My name is Sarana and I have come here to help you."

Instead of making funeral arrangements upon arrival, his

son the biochemist was told that his father had miraculously been cured. The son found no angelic precedent for a Sarana and the matter entered family lore. His father passed away in late 2004.

Five years later, during 2004, as the biochemist was working late, waiting for an experiment to run its course, he tried an Internet search on "Sarana" connected with the Tuatha de Danann, given what had emerged in the original DNA on Peter's case. The biochemist had only his father's phonetic version of the name of the angel who had rendered the miraculous cure. The web search revealed the name Sirona, which phonetically is very close to Sarana. Sirona was a Celtic Gaulish goddess of *healing*, fertility, and regeneration, and was often associated with medicinal springs. Her cult was spread widely through Europe, from Hungary to Brittany. Sirona was reckoned to translate as "divine star." In her appearances she often carried three eggs, obviously a fertility symbol. She also was described as having a snake coiled around her arm, with its head directed toward the egg.[186]

Anthropologist Jeremy Narby, in his provocative study *The Cosmic Serpent: DNA and the Origins of Knowledge*, notes that shamans often claim that the source of their impressive botanical knowledge is communications with the plant deities themselves, who are envisioned as coiled, intertwined snakes.

Snakes, fertility, eggs, forbidden knowledge, DNA—is this all a bizarre coincidence?[187]

Epilogue

T HE DNA PROFILING PROCEDURES WE UTILIZED IN THE
Peter Khoury "alien hair" investigation during 1998–
2001, while still having considerable utility, have
been overshadowed by more detailed and more complex
methods. Some are so sensitive that even a molecule of biological evidence is sufficient to undertake DNA tests. However, with that sort of increased sensitivity, contamination
and error factors loom as major concerns. Even with standard
PCR-based DNA profiling, considerable debate continues
about factors that might affect its accuracy.[188] Because forensic methodologies are changing rapidly, it is extremely important to keep up to date.

But at least the UFO research community now has a clear
choice. It can continue to languish in the uncertainty of unverified accounts and unsubstantiated theories, or it can at-

tempt to anchor these strange events with the focused attention of forensic and scientific investigations and research, in order to yield verifiable evidence on the reality of alien abductions, whatever that turns out to be. I choose the latter, and I hope others will share and support this direction. I believe every contact leaves traces, some hidden, others more obvious, which can yield evidence to illuminate our way through this extraordinary mystery.

While the Peter Khoury case confirms the utility of the DNA forensic approach, the real challenge ahead for researchers is to determine if these anomalies are both valid and significant. To do this, abduction researchers should cooperate in a testing program focused on DNA profiling. Testing a significant number of legitimate samples would provide an opportunity for validation of the unusual anomalies found to date. Additional results would add to the database of biological evidence of alleged alien specimens. Such a strategy could help us determine whether the aliens are a biological reality and if indeed any of them are visiting our planet and abducting humans.

On the evidence gathered so far, it seems that the alien abduction mystery strikes at the very heart of a profoundly significant issue—that of "humans" from elsewhere. In *Are We Alone?* theoretical physicist Paul Davies contemplated the possible existence of what he called "duplicate beings." He concludes by stating that "in most reasonable spatially infinite cosmological models with conservative assumptions, there are indeed an infinite number of duplicate beings." He then goes on to say: "If we were to discover extraterrestrial DNA that could be proved to be of independent origin, it would strike at the very heart of Darwinian evolutionary theory and the entire [currently dominant] scientific paradigm in which all teleology is decisively rejected."

Evolutionary paleobiologist Simon Conway Morris reaches much the same conclusion in *Life's Solution: Inevitable Humans in a Lonely Universe*. Using our best evidence for life, our own on planet Earth, to argue against the prevailing scientific evolutionary paradigm, he states that because of the ubiquity of evolutionary convergence, not only does life have an extraordinary propensity for navigating multiple pathways to precise biological solutions, but that it repeatedly reprises the *same evolutionary solution*, mediated powerfully by "the weirdest molecule in the Universe"—DNA. In short, Morris contends that on suitable planets out there the genetic tape of life will play out into more "inevitable humans." But, contends Morris, the rarity of Earth-like planets means that we are most likely living in a lonely universe.[189] Others argue, however, that life is everywhere, driven by a "life principle" that favors the spread of life through the universe.[190]

Paul Davies revisited the theme of extraterrestrial DNA in a way that took many people by surprise in 2004. He speculated in *New Scientist* that a more intelligent way to look for evidence of an extraterrestrial message, rather than SETI searches via radio astronomy, was to see if there were unusual extraterrestrial patterns in so-called "junk" DNA. Could evidence of alien intervention be hidden in our own human DNA? he wondered[191] The idea that there might be "alien messages in our genes" sounds like science fiction, but it begs serious consideration of whether we are the result of natural or artificial evolution. What if there are "non-Darwinian patterns" in our DNA? As it happens, my biochemist associates and I had already been looking for verifiable factors in "junk DNA," as well as unusual mutations (or polymorphisms) in our DNA, that might provide evidence of "non-Darwinian patterns" that are perhaps reflective of extraterrestrial influence.

As our search continues, I realize that when it comes to the subject of alien abductions, the sky is clearly not the limit. Whenever I have the opportunity to go to remote or different places, be it the outback of Australia, the Great Wall of China, the Great Lakes of the United States, the open moors of the Pennines in England, or other localities on this "pale blue dot"[192] called Earth, I always try to take in the night sky. It is in these sorts of places, where the sheer immensity and closeness of the night sky is very powerfully present, that the boundary between us and what lies out there often begins to recede in the face of contemplation. The Australian novelist Tim Winton captured it in his book title, *That Eye, The Sky,*[193] where visions of "otherworlds" are evoked and boundaries between the known and unknown become blurred. On these occasions I have the overwhelming sense that our little portion of the cosmos is far from unique, and that we as humans are far from unique as well. Life is out there.[194] It may not be all that surprising that one day we might find that some of it is like us in so many ways, even down to the basics, DNA itself. And indeed, some of that humanlike life may already be here, making very strange intrusions into our world.

PHASE ONE: HAIR SHAFT ANALYSIS—
STRANGE EVIDENCE

What follows is the detailed technical report prepared by the lead biochemist of the team that facilitated this work.

Mitochondrial DNA Sequence Analysis of a Shed Hair from an Alien Abduction Case

by the Anomaly Physical Evidence Group (APEG), April 1999

Introduction

A great controversy currently exists as to whether the many reported "alien abduction" cases worldwide might be physically real or else purely psychological phenomena. Thus, those people who believe in the UFO scenario tend to accept the investigations of authors such as B. Hopkins[195] or J. Mack[196], who have provided circumstantial evidence for the abduction of humans in terms of: (a) landing-site traces, (b) prominent "biopsy" scars, (c) alleged "implants" (or artifacts ostensibly removed from abducted people), and (d) testimonies from hundreds of supposedly abducted individuals, whether by conscious recall or under hypnosis. However, those people who do not believe in the UFO scenario (and this would include the majority of professional scientists) tend not to accept the current evidence for alien abductions, considering it to be too weak and circumstantial to justify such a major shift in their way of perceiving the world.

Indeed, these supposed abductions leave surprisingly little evidence that could be tested in a scientific laboratory, according to standard procedures; and hence, if real, they might be intentionally covert in nature.

The detailed analysis of any reliable evidence from an alien abduction case might therefore be of great scientific value, in order to assess the reality of this phenomenon, as well as to identify the biological nature of such visitors.

A Detailed Case Study

One such piece of physical evidence was obtained recently from an alien abduction case in Sydney, Australia, in 1992, when supposedly two near-human females appeared suddenly to a young man in daylight in his bedroom. This same individual had previously in 1988 experienced an event in which several aliens had entered his house and had ostensibly obtained "biopsy" material, while his family in the next room was put to sleep. He reported the experience to a number of people, including local investigator and researcher B. Chalker[197]. After that event, a fresh "punch-biopsy" scar was noted on the shin of the young man's lower leg, and a photographic record of the scar was made.

Now, for this later event in 1992, the young man reported that two near-human females suddenly appeared on his bed, while no one else was in the house, and attempted to engage him in an apparent sexual embrace. One girl was described as tall (6 feet), of fair appearance with light blond hair; while the other girl was of medium height (about 5 feet 6 inches), of Asian appearance with dark hair. Both supposedly showed near-human female bodies, but displayed unusual racial characteristics in their faces, which in the case of the blond female was long and narrow, and also in their eyes, which were exceptionally large.

Due to the shock of such contact the young man struggled with the tall female, and after what seemed a few moments while the young man was distracted with a coughing fit, he found that he was alone. Still it seems that the tall blond female left clear evidence of her presence, in the form of blond head hair of length 15 cm, which was wrapped tightly under the young man's foreskin.

This hair was immediately retrieved by the young man, stored in the dark in a sealed plastic bag, and not touched by anyone else prior to its scientific investigation for DNA evidence in 1998. The young man did show the sample in its sealed bag to a few people between 1993 and 1998, but he is certain that no one, other than himself, handled the hair directly. In any event, no evidence of contaminated DNA from the outside of the hair was found.

Forensic Investigation
In early 1998, the investigator B. Chalker provided such hair to the Anomaly Physical Evidence Group, whose members as professional scientists choose to remain anonymous at this time. A standard forensic investigation was then carried out on the hair shaft, using well-established protocols[198], and also on control hair samples from the young man and his wife (married in 1990). The goal of such analysis is to establish a precise DNA base sequence of mitochondrial hypervariable region I, spanning nucleotides 16,000 to 16,400 of the circular mitochondrial DNA. Such DNA is present in hundreds of copies within each human cell, and thereby acts as an easily amplified genetic marker for the polymerase chain reaction (PCR), even in moderately degraded samples. [See MITOMAP on the Internet for more information: http://infinity.gen.emory.edu/mitomap.html for Mitomap v.3.1—A human mitochondrial genome database, Center for Molecular Medicine, Emory University, Atlanta, GA, USA, 1999.] Any sequence variation within that hypervariable region 16,000 to 16,400 then provides a "microscopic genetic fingerprint" of the individual involved, which is used commonly as forensic evidence in many criminal investigations.

In order to extract fragments of mitochondrial DNA from the hair of the tall, blond alien female, a 2 cm piece of hair shaft (located just above the root) was excised and transferred to a sterile Eppendorf tube. Similar pieces were excised also from both control hair samples, as obtained from the young man and his wife, in the presence of the investigator B. Chalker. These three hairs could be easily distinguished by their appearance under a microscope, since the hair from the tall blond female was extremely thin and almost clear (very little melanin); whereas the hair from the young man was of normal thickness and black; while hair from his wife was of normal thickness and brown. In fact, that hair from the tall blond alien was found to be almost invisible to the naked eye on a glass surface, due to its unusual optical clarity; and hence it could only be handled under reflected light. Further investigation of this thin, almost-clear hair by high resolution darkfield microscopy showed it to lie at the lower end of normal human hair thickness; and also to show a pronounced "mosaic" structure, perhaps due to the near absence of melanin.

Procedure

The three hair samples to be tested: (a) tall blond alien, (b) the young man, and (c) his wife, were washed twice with PCR-quality water, twice with 70% ethanol, then once with extraction buffer (7 M guanidinium hydrochloride, 100 mM Tris, pH 7.0, 1% Triton X100 and 5 mM EDTA) for 1 hour at 20°C, to remove any contaminating DNA from the outside of the shaft. Next, each washed hair was extracted in the same guanidinium-Triton buffer plus 10 mM dithiothreitol (DTT) for 2 days at 50°C. Finally, each hair sample was fragmented using a sterile disposable pestle for 1.5 ml Eppendorf tubes, then subjected to several cycles of boiling and freezing on dry ice, in order to release the DNA from its protein matrix.

Each fully-treated hair solution was then extracted twice using phenol/chloroform/isoamyl alcohol (25/24/1) to remove organic impurities, especially melanin. The purified DNA was then supplemented with 2 micrograms of UV-sterilized glycogen as a carrier, and precipitated with ethanol. Each DNA pellet was next resuspended in 10 mM Tris, 0.1 mM EDTA and used in small aliquots for PCR amplification.

No washes of any hair, prior to extraction with guanidinium-Triton and DTT, produced any detectable mitochondrial DNA in subsequent PCR steps. Other attempts to extract DNA from pieces of that hair far from the root were unsuccessful.

In order to amplify mitochondrial hypervariable region I from a broad mixture of DNA fragments, two oligonucleotide DNA primers were synthesized first over mitochondrial locations 15,993–16,022 (upper strand) or 16,401–16,430 (lower strand), because the nucleotides in those locations are highly conserved among all primates, including chimpanzee, gorilla, orang-utan and man. Next, two more primers were synthesised within the hypervariable region itself, so that one might amplify the whole DNA of 16,023–16,400 in two overlapping fragments (mitochondrial DNA from a hair shaft is often degraded, so that chains longer than 300 bases become difficult to find). Those two internal DNA primers extend over locations 16,190–16,209 (upper strand) or 16,271–16,294 (lower strand) within the hypervariable region itself.

Thus, when primers 15,993–16,022 (upper) and 16,271–16,294 (lower) are used as a pair in PCR, they will yield a mitochondrial DNA fragment spanning the left-hand side of the hypervariable region, as a product of size 302 base pairs. Similarly, when primers 16,190–16,209 (upper) and 16,401–16,430 (lower) are used as a pair, they will yield a 241 bp product spanning the right-hand side of the hypervariable region (see MITOMAP on the Internet for details, ibid).

Each PCR reaction was carried out under standard conditions using 1.5 mM magnesium chloride, and for 30 to 36 cycles of denaturing at 95°C, extending at 72°C, and annealing at 57°C. All reactions were checked by agarose gel electrophoresis after just 30 cycles, to see whether the amplification had proceeded efficiently; and then run for another 6 cycles if only small amounts of product were observed. The background contamination of control samples not containing any hair (by other DNA in the laboratory) was never more than 10% of the product.

Once PCR products from both the left and right-hand sides of the hypervariable region were obtained, in quantities of at least 100 nanograms, each DNA band was purified by excision from an ethidium bromide–stained agarose gel, following several hours of electrophoresis. Next, each excised band was treated with high salt and silica beads, to extract the DNA. Finally, each amplified DNA fragment of size 302 or 241 base pairs was trimmed with Pfu polymerase to create blunt ends for cloning, then treated with polynucleotide kinase to add 5'-phosphates.

Results

Four PCR products were obtained in total: two from the hair of the tall blond female (both left and right), two from the young man who collected the hair (both left and right), but none from his wife whose hair may have been chemically treated, so as to make recovery of DNA difficult (and who never came into direct contact with the alien hair).

These four amplified DNA fragments, two from the blond alien hair and two from the young man, were then cloned into a commonly-used plasmid vector. Four full libraries of clones were obtained, by transformation of the four ligation reactions into *E. coli*, followed by multiple preparations of plasmid DNA on a small scale.

Roughly 6–12 clones were obtained from each amplified DNA fragment as a fairly substantial library, by which to assess whether the amplified DNA might be pure and authentic, or else contain impurities due to contaminating DNA. Screening of these libraries using the dideoxy method for just one nucleotide showed that all four libraries contained pure and homogenous DNA, without any sequence impurities to a 90% level. Some clones were of reduced length, but these turned out to be end-deletions as produced in the PCR step or subsequent manipulation.

Full sequence analysis of all four bases G, A, T and C using the dideoxy method (Sequenase Version 2.0) showed that all clones from the young man's hair matched closely the human consensus from MITOMAP, which is European in nature. Thus, clones from his hair did not show any systematic deviation from the consensus in any of 380 locations spanning the whole hypervariable region 16,023–16,400, apart from a few common C-to-T or A-to-G heteroplasmies at 16,189, 16,270, 16,312 or 16,362.

By way of contrast, all clones as amplified from the hair of the tall blond alien female show five consistent substitutions from the human consensus. These are all C/T or A/G transitions, and are located at 16,108 (C to T), 16,129 (G to A), 16,162 (A to G), 16,172 (T to C) and 16,304 (T to C). Three of those transitions at 16,129, 16,172 and 16,304 seem fairly common among human racial types, whereas the two at 16,108 and 16,162 seem quite rare.

Thus, mitochondrial DNA analysis of the hair shaft from a reportedly tall, blond alien female shows that she is biologically close to normal human genetics, but of an unusual racial type. For comparison, Neanderthal man differs from human at 27 locations in the same DNA, while chimpanzee differs from human at 55[199]. One might expect from the physical description of that tall blond female, as well as from the possible sexual nature of the event, that such a tall blond female might represent just some strange human racial type. One might predict further that her DNA should match closely that of racial types in Finland, Iceland, or Scandinavia, given the long, thin blond hair as direct evidence, plus her tall stature and fair skin as eyewitness testimony; but as we shall see below, that seems not to be the case.

Because of the subsequent clarification of the nature of the anomalous hair DNA sample, additional controls from blood samples were obtained from the young man, his wife, and a Chinese man who spent some time in the same room as the alien hair (but who never contacted it directly).[200] From these, mitochondrial DNA was amplified and analysed as before in the region 16,023–16,270. The results show that all three control samples lie very close to a modern human consensus, with few if any substitutions as seen for the alien hair.

Thus, the Chinese man shows a single substitution at 16,223 (C to T), typical of most Asians; while the young man's wife shows no substitutions at all. Finally, the young man shows a single substitution at 16,249 (T to C); a region near 16,190 remains unclear. The heteroplasmies seen for DNA taken from the young man's hair, in no more than one out of 6–12 clones, may therefore represent just occasional mutations in an aged hair, since they are not found in blood.

In summary, the blond alien hair provides for a strange and unusual DNA sequence, showing five consistent substitutions from a human consensus (present in all clones), which could not easily have come from anyone else in the Sydney area except by the rarest of chances; is not apparently due to any sort of laboratory contamination; and is found only in a few other people throughout the whole world (see below).

Comparison of Present Results with Those Reported Elsewhere

A detailed survey of the vast literature on sequence variation in hypervariable region I of mitochondrial DNA (see MITOMAP, ibid.) revealed that only 4 persons from this entire literature, consisting of tens of thousands of individuals, contain the extremely rare C-to-T substitution at 16,108. Quite surprisingly, those four persons who contain a C-to-T substitution at 16,108, also contain all four of the other alien-type substitutions at 16,129, 16,162, 16,172 and 16,304; yet show almost no other changes in the entire hypervariable region of 380 base pairs (one person differs by a single base). Hence, we may conclude with high probability, that those four human persons and the tall blond alien girl share a common maternal ancestor, sometime in the past 2,000–10,000 years, given known rates of substitution in mitochondrial DNA. Indeed, a perfect 5/5 match between the tall blond alien and those four persons indicates that little if any random substitution have occurred in the intervening period.

Who might those four persons be, who seem to share a distant maternal ancestor with the tall blond female, who left her hair with a young man in Sydney in 1992? It turns out that all four are of the Mongoloid Chinese racial type, with presumably Asian appearance as well as dark black hair[201, 202]. One was included as part of a small group from China[201], while the other three were found as just 4% in a large group of Taiwanese[202] (see D84952, D84956 and D84985 from the DNA Data Bank of Japan).

What implications might these comparisons have, for possible authenticity of the alien hair sample as collected by the young man in Sydney in 1992? While it would not be impossible for him to have had sexual contact with some fair-skinned, nearly-albino female from the Sydney area, such an explanation is ruled out by the DNA evidence, which fits only a Chinese Mongoloid as a donor of the hair. Furthermore, while it might be possible to find a few Chinese in Sydney with the same DNA as seen in just 4% of Taiwanese women, it would not be plausible to find a Chinese woman here with thin, almost clear hair, having the same rare DNA. Finally, that thin blond hair could not plausibly represent a chemically-bleached Chinese (including the root), because then its DNA could not easily have been extracted.

The most probable donor of the hair must therefore be as the young man claims: a tall blond female who does not need much colour in her hair or skin, as a form of protection against the sun, perhaps because she does not require it. Could this young man really have provided by chance, a hair sample which contains DNA from one of the rarest human lineages known (family C2 of ref. 6), that lies further from the mainstream than any other except for African Pygmies and aboriginals (family C1)? Note that a few other persons from China, Japan or Korea show a partial 4/5 match to the blond alien DNA, but none to date show the rare change at 16,108.

General Implications and a Need for More Research

Studies of mitochondrial DNA among various human types have led to disturbing conclusions, if one accepts the Darwinian theory which suggests that we evolved gradually from apes, by random mutation followed by natural selection. For example, humans appear to be far less genetically diverse than other species such as chimpanzee, which suggests a recent "bottleneck" in our origin[203]. Also modern Caucasians and Asians differ greatly from the lower primates in terms of genes for Rh factor (a determinant of fertility) and DNA sequences on the Y chromosome (among others). Could these large changes of DNA structure have come about just through random mutation and drift? Or might the modern human population be a recent introduction to Earth from elsewhere, say 30,000 years ago when the Neanderthals went into rapid decline?

If the tall blond alien and a few Chinese Mongoloids share a common maternal ancestor within recent historical times, could it have been in the Yellow River region of China, where civilization began around 3000 B.C.? Or perhaps further west in Taklimakan, where ancient fair-skinned mummies have been found in an excellent state of preservation? [See www.pbs.org/wgbh/nova/china-mum/taklamakan.html; or "Mummies of the Taklamakan" TV documentary, 1999. See also "Riddle of the Sands" by Peter Martin, *The Australian* magazine, March 13–14, 1999; *The Mummies of Urumchi* by Elizabeth Wayland Barber, Macmillan, 1999.]

We cannot know the answers to these questions without much more open-minded, well funded scientific research. Today most professional scientists deny that there could be any humans elsewhere with greater technical capability than our own; deny that there could exist any humans co-fertile with us from elsewhere; and deny that ancient people could have been superior to ourselves in the late 20th century.

Rather than adhering inflexibly to these conservative views, we feel that the present study suggests:

(1) there do exist humans other than our kind elsewhere, who have technological capacities beyond our own;

(2) these other humans may occasionally visit us for purposes of reproduction (or other);

(3) a few of the exotic humans as represented by that tall blond female may have lived in China long ago, and left descendants. Indeed, the second girl seen in Sydney 1992 was reported to be Asian in appearance, even though DNA evidence to that effect was not obtained until 1998. Multiple washes of the blonde hair prior to DNA extraction showed no contaminated DNA on the outside and hence the DNA we have obtained must have come from the inside of the hair, presumably from the blond female herself.

Great progress could be made if both UFO and alien abduction studies were accepted within the scientific community as viable research, so as to deserve open discussion and funding of a high priority. Scientists of all kinds could then work with competent anomaly investigators as we have done here, to obtain samples for research; while the primary providers of such evidence need not be scorned, but could be treated as favourably as say Schliemann who found Troy, or some archaeologist who finds a novel human skeleton. Why should any science be "forbidden"?

Acknowledgements

This research was conducted on private funds ($5000) and in a private laboratory, without outside support. The APEG is a professional scientific group with no outside ties. Its members wish to remain anonymous for now. Copies of a video tape on which the alien hair appears by darkfield microscopy, as well as copies of the alien DNA sequences which were cloned into plasmids, have been made and may be made available to serious researchers on request and determination of appropriateness.

PHASE TWO: HAIR ROOT ANALYSIS— VERY STRANGE EVIDENCE

The principle biochemical scientist working with the APEG compiled the following report.

Genetic Analysis of a Hair Root from a Reportedly-Alien Blond Female: Mitochondrial and Nuclear DNA
by the Anomaly Physical Evidence Group, May 2000

Introduction

As we enter the 21st century, on this little planet at the outer edge of a spiral galaxy, we hear almost daily some new report of strange, exotic aircraft flying through our skies. Those craft have been seen clearly (and sometimes photographed) by ordinary citizens, military personnel or police, on many occasions over the past 50 years. They have left direct physical evidence in terms of landing traces, radar images, and inhibitory effects on power supplies, in dozens of well-documented cases. Many serious workers believe that such craft may be occupied by visitors from other worlds, because they display a wide range of advanced technologies which exceed greatly what we are capable of today.

However, relatively few people have ever come into direct contact with a living being not from Earth, and even fewer have been able to provide plausible evidence of a biological close encounter. Thus the genetic analysis of a shed hair from a reportedly-alien blond female, as reported here, may be of some interest as the first investigation of its kind.

Recovery of the Sample

This strange hair was collected from a suburb of Sydney, Australia, in July of 1992, following an apparent close encounter between a young man, Peter Khoury, and two almost-human females in early daylight in his bedroom. One visitor was reportedly tall and blond with thin sparse hair, while the other was of medium height and Asian with dark black hair (*IUR* Spring 1999, pp. 3–31). A tall blond alien has been described also in many other cases: see for example *Men in Black* (1997) by Jenny Randles, or *Fire in the Sky* (1978) by Travis Walton.

Although certain details of Peter's 1992 encounter remain unclear, a thin blond hair was found afterward in a location almost impossible to miss, since it was wrapped tightly under the young man's foreskin. Hence this particular sample could not have been recovered erroneously from a pillowcase or bed sheet, as a normal human hair. Furthermore, its clear thin morphology makes it quite unlike any normal human hair, both in color and in thickness. Finally, Peter has seemed continually honest and consistent in his reporting of this episode over the past 8 years. He would have no easy way of finding a long, thin colourless hair in Sydney; nor does he have any knowledge of hair forensics.

The Khoury family comes from a priestly, Christian Maronite background in Lebanon. Peter's first encounter with a non-Earthly craft was in a small coastal town of Lebanon in the summer of 1971. At age 7, he and eight other children were playing on a rooftop, onto which Peter walked last through a heavy door. He then saw all his friends "frozen" like statues in front of him, while a silent egg-shaped craft hovered above. All eight children and Peter later found themselves on the ground after some time had elapsed, with no memory of the intervening period.

Much later in Australia in July of 1988, Peter was asked by his brother to go to his room in his father's house, so that his brother could watch TV in the main room. Immediately after entering the room, Peter felt a "pins and needles" paralysis, and saw a variety of robed non-human aliens in front of him. Two 7-foot humanoids informed him telepathically to "stay calm, do not be afraid, it will be like the last time"; and then inserted a needle into his head which caused him to lose consciousness. Meanwhile, his father and brother were put to sleep in the main room for many hours.

Still later in a different house in Sydney in July of 1992, Peter had a daylight encounter with two almost-human females as noted above, which resulted in the recovery of a clear, thin head hair of length 15 cm. Another brief encounter with non-human aliens was noted in November of 1996, when he was apparently carried late at night through a mirror by teleportation. All four episodes from 1971 to 1996 show a long-term periodicity of 4.25 years, which incidentally is twice that of Mars orbit (2.1 years).

Physical Evidence for Multiple Encounters

What evidence do we have to think that Peter might be telling the truth? Could he somehow be imagining the events of 1988, 1992 and 1996, due to a shock from his childhood encounter of 1971? If so, he would have had to inflict several large, surgical-like scars on himself, without pain and without any instruments to do so. Thus, all four episodes seem to have left direct physical evidence on Peter's lower right leg, in the form of four punch-biopsies of diameter 8 mm.

Such large, round, hairless scars could not easily be self-inflicted, for two reasons. First, punch biopsies are seldom performed by Earth doctors due to the intense pain of taking a core sample, plus the need for tight stitching afterward, and a long time of 2–4 weeks to heal. By contrast, all of Peter's scars remain unstitched, and one healed itself in 6 hours while his wife watched. (It would be a boon for medical science to elucidate the mechanism of such rapid wound healing.) Second, punch biopsies when performed here on animals are carried out using a simple cylindrical tube-like blade, that is inserted directly in or out. By contrast, Peter's punch biopsies (like those of many abductees) show fine-structure around the outer edges, where 6–8 thin wires have been inserted to prevent torsion, while twisting the central tissue as for a corkscrew.

Similar punch-biopsies have been noted also in other cases: see *Intruders* (1987) by Budd Hopkins, *Abduction* (1994) by John Mack, or *The Aliens and the Scalpel* (1998) by Roger Leir; but have not yet been explained by conventional means.

Preliminary Analysis of the Upper Hair Shaft

Given so many supporting data, which suggest that Peter might be testifying correctly about his multiple encounters, and also about collection of a strange hair sample, we decided to make DNA analysis of that hair a high priority. We therefore assembled a team of highly competent (if anonymous) biological scientists to carry out such work, with the substantial cost paid fully from the investigators' own private funds.

First, high-resolution photographs of the blond "alien" hair by darkfield microscopy showed a pronounced mosaic structure, owing to the near-absence of melanin, and confirmed that it was a variant form of human hair (*IUR* Spring 1999, pp. 3–31).

Next, in order to make sure that our DNA methods were working well (before studying the root), we analysed a 2 cm portion of thin blond hair shaft, which was cut from just above the root. That hard hair shaft required extensive grinding to release its DNA; no contaminating DNA was found on any outer part. Studies by PCR (polymerase chain reaction) were then used to deduce the identity of all DNA released from cells on the inside. Those studies revealed an unusual mitochondrial DNA of Chinese genetics, with a rare mutation 16,108 C-to-T, plus four other mutations 16,129, 16,162, 16,172, 16,304 in hypervariable region I (HVRI).

One could perhaps collect the black hairs from 100 Chinese in Sydney, and find one example of such rare DNA. Yet since the blond "alien" hair never came into contact with any local Chinese, and since it is optically-clear rather than black, that scenario seems unlikely. Nor does laboratory contamination provide a plausible explanation, since careful controls were performed at every step. Alternatively, we might have amplified DNA from the second alien who visited Peter in 1992, if that Asian-like female tied a head hair from her blond friend about the young man's foreskin. Yet why could we not recover Chinese DNA from the outside of that hair, if the Asian female handled it?

Our preliminary study therefore left two anomalies:

How could Chinese mitochondrial DNA of a rare type be found within any hair which is optically clear?

How could such rare DNA be collected from the foreskin of Peter Khoury, unless it was left there by an alien female as he says?

In order to resolve these anomalies, we undertook further investigation of soft root tissue from the same hair, which generally contains more DNA.

Methods for DNA Extraction and Analysis by PCR

Our methods for DNA extraction remain almost the same, except that we omitted treatment of all hair-root extracts with phenol/chloroform, because they seem free of any melanin which might inhibit PCR. Also, we first extracted soft tissue from around the root by a 20°C wash for 30 minutes in guanidinium buffer; then we extracted the residual hair shaft by a 55°C wash for 2 days, followed by freeze/thaw and grinding. Thus, the soft root–tissue and hard hair–shaft were extracted separately as independent samples for PCR.

Our PCR methods also remain the same, where both the left and right-hand sides of mitochondrial HVRI (hypervariable region I, nucleotides 16,020–16,400) were amplified for 37 or 33 cycles respectively; while one small nuclear region which codes for CCR5 was amplified for 40 cycles. Other nuclear regions gave no amplification, because the 8-year-old hair DNA is apparently too degraded to do so. Recall that mitochondrial DNA is present in any cell at a level 200–500 times that of nuclear, and hence is the more easily detected in an aged sample.

A putative "yowie" hair[204] was tested in parallel with the blond "alien" hair, but yielded no ape-like DNA. Still, that yowie hair serves as a good control for possible laboratory contamination, since it was amplified with the same solutions and at the same time as the blond hair. Only very-weakly amplified products, almost identical to a European consensus, were obtained from the "yowie" hair. Hence, there is apparently little contamination by mitochondrial DNA from local sources in our protocol, except for consensus European at a very-low level as noted in Table I. A PCR test for nuclear CCR5 also showed no contamination from any local source.

New Results on Mitochondrial DNA from the Blond "Alien" Hair Root

Our new results on the blond "alien" hair may be summarized as follows: its mitochondrial DNA appears to be of two kinds, depending on whether we analyse the hard hair shaft or else the soft root.

Thus, from the lower hair shaft we obtain again the same Chinese mitochondrial DNA, which shows a characteristic substitution 16,304 T-to-C. But from soft root tissue, we obtain a novel Basque-Gaelic mitochondrial DNA, which shows a rare substitution16,255 G-to-A, along with several other characteristic substitutions 16,223 C-to-T, 16,278 C-to-T, and 16,294 C-to-T (Piercy, R., et al., *International Journal of Legal Medicine* 106, 85–90, 1993; Bertranpetit, J., et al., *Annals of Human Genetics* 59, 63–81, 1995; Corte-Real, H., et al., *Annals of Human Genetics* 60, 331–358, 1996).

A detailed account of our new results is listed in Table I. These new data refer only to the right-hand side of HVRI (16,210–16,400), because DNA from the long left-hand side (16,020–16,270) could not be obtained, perhaps due to degradation. Our previous study of an upper hair shaft shows the same Chinese mitochondrial DNA, but with five substitutions at 16,108, 16,129, 16,162, 16,172 and 16,304 as measured in both left and right-hand regions.

TABLE I. NEW RESULTS ON MITOCHONDRIAL DNA FROM A BLOND "ALIEN" HAIR ROOT

Sample	Mitochondrial sequence variation	Number of clones
Outside root-tissue	16,223, 16,255, 16,278, 16,294	6/7
Lower hair-shaft	16,304	3/6
Upper hair-shaft	16,108, 16,129, 16,162, 16, 172 16, 304	6/6

NOTES TO TABLE I

All base changes are A/G or C/T in standard numbering. In each case, a majority of clones show the characteristic sequence, while others show very close to a European consensus. One clone shows 16,249 as for Peter Khoury, and may be due to his handling of the hair in 1992. The outside root tissue and lower hair shaft were measured only from 16,210–16,400, while the upper hair shaft was measured fully from 16,020–16,400.

By comparison with all other human sequences which have been determined (Mitochondrial DNA Concordance, http://shelob.bioanth.cam

.ac.uk/mtDNA), we find that the four substitutions found here for a blond "alien" hair root, definitely place that female within the general range of Basque-Gaelic genetics. Thus, two of the four substitutions 16,223 and 16,278 are found together in 3% of Caucasians from the U.K. or northern Spain (Piercy et al., 1993), or in roughly 10% of the blue-eyed, blond Caucasians who inhabit northern Scotland and the Orkney Islands.

Examining next only those people who contain both 16,223 and 16,278, we find that even fewer (1% in the U.K. or 0.5% in northern Spain) contain an additional substitution at 16,255. Finally, no one in the current database contains a fourth substitution at 16,294, in combination with the other three. Yet 16,294 can be found along with 16,223 or 16,278 in some individuals.

Recent studies of the Y chromosome (Hill, E., et al., *Nature* 404, 351, 2000) enable us to pinpoint the location of our blond "alien" genetics more closely. Thus, it seems that the western Irish and Basque share a rare Y-chromosome haplotype known as "hg 1," not found commonly anywhere else in the world. The geographical distribution of "hg 1" seems to match well the geographical distribution of mitochondrial DNA from our blond "alien" hair; which suggests that both kinds of genetic variation may have spread from western Ireland, by population expansion into Scotland and northern Spain.

Thus, the rare Basque-Gaelic DNA as found within soft root tissue taken from this "alien" hair is clearly consistent with a blond, blue-eyed racial type. How could alien DNA enter the human race, as a trace component of its total gene pool? We will discuss below the possible introduction of alien DNA into ancient Ireland by interbreeding, as well as the anomalous detection of Chinese mitochondrial DNA within a hard shaft of the same sample.

PCR Test for Nuclear CCR5

Let us turn next to the possibility of recovering some information about single-copy nuclear DNA from our "alien" hair. Most tests for nuclear genes gave no amplification, by comparison with human controls. However, one test for a small deletion known as delta-CCR5 gave a promising smear of bands, when tested over a large size range of 158–190 bp (base pairs). Therefore, we repeated the test for delta-CCR5 using a small size range of 58–90 bp, with primers.

The gene CCR5 is very important in human genetics, because it confers immunity to AIDS-HIV when deleted (HIV virus enters the cell through a pore formed by CCR5 protein). The mutation delta-CCR5 may confer resistance to other viruses as well, for example smallpox or myxomatosis. This 32-bp deletion is quite rare, and found mainly in people of northeast European or Jewish descent, with an estimated origin just 5000 years ago.

PCR results for delta-CCR5 using the blond "alien" hair show two human controls on the right show either a large band of size 90 bp (lane 4), which indicates normal CCR5 on both chromosomes; or else a large band plus small band of sizes 90 or 58 bp respectively (lane 5), which indicates one normal CCR5 and one deleted. The vast majority of modern humans show a single band of 90 bp, while a few people show two bands of 90 or 58 bp. Very few people (less than 1%) from Scandinavia, the U.K., northern Spain or Ashkenazi Jews show a single band of 58 bp, which indicates deleted CCR5 on both chromosomes. In order to be viral resistant, a person must contain two copies of deleted CCR5, so that no protein will be available to let viruses enter the cell.

Now on the left-hand side of the PCR results every PCR sample from the blond "alien" hair shows two small bands of size 58 or 50 bp: a preliminary water-wash (lane 1, weak), soft root-tissue (lane 2), or hard hair-shaft (lane 3). Compare these bands with lane 4 (two normal CCR5), lane 5 (one intact plus one deleted CCR5), or lane 6 (no DNA). It appears then, that the blond alien female might contain two deletions for CCR5, making her viral resistant. At the very least, our samples seem to show no contamination by local human DNA, which would produce a band of 90 bp. We tested our primers on other samples, but none gave two short bands as shown in the CCR5 PCR results. Nor did any sample show short non-specific products as primer-dimer. [See photo insert for CCR5 DNA sequence.]

Cloning and Sequencing of CCR5 from the Blond "Alien" Hair

Detailed cloning and sequencing of PCR products was therefore undertaken, in order to check the specificity of our result for CCR5, as well as to determine the sequence amplified between our two primers. These results are summarized, where bases which were amplified are shown in capital letters, whereas bases which were not amplified are shown in small letters. Out of 24 clones examined, all short ones show 2 or 0 bp between the primers, typically sequence AG or nothing. Alternatively, all long ones show 7–12 bp between the primers.

The sizes of those two blond "alien" bands therefore match well, the expected 10 bp for human delta-CCR5 on one chromosome, plus a smaller 0–2 bp on the other chromosome. Such sequencing results become ambiguous, however, when examined in closer detail.

Thus, human delta-CCR5 should show a sequence CATACA-TTAA between the primers, whereas clones from the blond female show various CCR5 sequence, but not the full CATACA-TTAA in any one clone. Hence an alternative interpretation might be that those two small bands arise by recombination or jumping, in the early part of a PCR reaction on an aged sample. This result remains ambiguous, because we have not yet been able to reproduce a small two-band pattern using any other sample; and it indicates the need for much more research with ample time and funding.

The CCR5 PCR results:
PCR amplication products of the nuclear CCR5 gene, as applied to a 3% agarose gel and stained with ethidium bromide.

    lane 1: blond alien hair, water-wash;

    lane 2: blond alien hair, soft root-tissue;

    lane 3: blond alien hair, hard-shaft;

    lane 4: normal human control, two intact CCR5 genes;

    lane 5: normal human control, one intact (90 bp) and one deleted (58 bp) CCR5 gene;

    lane 6: no added DNA. Two outside lanes show size markers, the smallest of which is 75 bp.

Summary and Discussion
Here we have performed the first detailed genetic analysis of a tissue sample from a plausibly alien source, namely the hair root of a reportedly-alien blond female. One may be assured that the scientists who have conducted this study are of the highest quality, and have

published many papers in top journals on various aspects of DNA and its analysis by PCR. We have been limited mainly by the poor quality of sample: an 8-year-old hair root, which yields relatively little mitochondrial DNA and almost no nuclear. Given a trace of fresh blood from that blond alien, or even a fresh hair root, one might have been able to analyze her entire genome, and make a genomic library available to biologists worldwide. But that promising scenario will never be the case, so long as sample collection and scientific research are hampered for political reasons.

One recalls how prominent academics in Galileo's day refused to look through his telescope, or refused to watch whether two balls of unequal mass might fall at equal rates, because it would upset their long-held Aristotilean philosophies: see *Galileo's Daughter* (1999) by Dava Sobel. Is that any different from today, where followers of Einsteinian relativity refuse to consider numerous well-documented UFO reports, which might imply anti-gravity propulsion, or travel faster than light? Or where followers of Darwin refuse to consider reports of almost-human species from space, which might imply that man did not evolve from the apes? See "Life Beyond Earth" by Joel Achenbach, *National Geographic,* January 2000, pp. 24–51, for a summary of modern academic views. See also Achenbach's book *Captured by Aliens: The Search for Life and Truth in a Very Large Universe* (1999). (Interestingly, a serious UFO report is given on the same issue—"Rediscovering America" by Tim Brookes, pages 69 and 72—a "ghost light" style report.)

The blond "alien" hair as studied here does not prove unequivocally that other near-human species exist, but it does raise important questions for further research. First, given the odd circumstances of recovery of such a hair, one is entitled to wonder why any tall, blond alien would tie her hair tightly around a young man's foreskin, so that it would be found and studied in detail? And why might she have chosen that particular young man over all others for such a close encounter?

Odd Recovery of a Thin Colourless Hair

Peter Khoury shows an intriguing family history, since he comes from a priestly line of the ancient Maronite religion in Lebanon, where Khoury means priest. Also, he grew up near the famous Temple of Baalbek, approximately 50 miles from where he had his first close encounter in 1971. Paranormal encounters with children seem surprisingly common, based on much anecdotal evidence from family or friends of APEG, who report contacts as early as 1957 in Australia or 1963 in Alexandria (often with punch-biopsy scars to support their case).

Twenty years later, after moving to Australia, Peter reported that several non-human aliens entered his bedroom in his father's house one night in 1988. To Peter, they were gentle looking and of both sexes. Those aliens performed various medical tests on him, such as drilling a tiny hole in his skull, or taking a punch biopsy from his lower leg. Four years later in 1992, two almost-human females appeared in Peter's bedroom in early daylight. Those women, whom he thought at first were normal human intruders, may have taken a sperm sample for breeding, based on the soreness felt afterwards, and they left their mark by wrapping a long head-hair from the tall blonde tightly under his foreskin.

Seen from an open-minded perspective, the most logical scenario would seem to be that higher creatures not from Earth have been following Peter's genetics for many years, or perhaps many generations; and that the almost-human females he met in 1992 may have taken advantage of such information (e.g., a planetary database of human genetics), to select for certain favourable traits in their own reproduction. We have no reliable information about where the non-human or almost-human aliens come from, nor how they entered Peter's house in 1992, nor how they put his father and brother to sleep in 1988. Yet the clear implication is that humanoid or near-human life forms can be found elsewhere in our galaxy; and that at least some of them are close enough to us genetically to interbreed without any fertility barrier. Indeed, the tall blond alien reported by Peter Khoury in 1992, bears a close resemblance to the short blond alien reported by Antonio Villas-Boas in Brazil 1957.

Could Peter have faked recovery of the blond "alien" hair in order to support his case? If so, he would have had to fake four punch-biopsy scars on his lower leg, as seen in many abductees; and he would have had to provide an anomalous hair which is as thin and clear as nylon fishing line. In fact, Peter was not even sure that his sample was actually a hair, until investigated in 1998 by darkfield microscopy for its mosaic structure (*IUR* Spring 1999). Nor has Peter at any point received favourable publicity or money for his story.

Ancient Interbreeding Between Blond "Aliens" and Local Humans?

In support of Peter's account, we have found over several years of careful study, that the blond "alien" hair contains two kinds of rare mitochondrial DNA, neither of which could be recovered easily from anyone in Sydney. First, soft scalp-tissue from the root contains a rare Basque-Gaelic mitochondrial DNA, with sequence changes at 16,223, 16,255, 16,278 and 16,294 relative to a human consensus. Such DNA has not been found anywhere else in the world, apart from 0.5% of Caucasians in northern Spain and the U.K. A double substitution 16,223–16,278 is common among blond, blue-eyed people of northern Scotland and the Orkney Islands, while 16,255 and 16,294 are less common substitutions within the same genotype. Finally, this particular mitochondrial variation seems to correlate well with haplotype "hg 1" for the Y chromosome, which shows a high concentration in western Ireland and northern Spain.

If our results are true, they would imply that distant ancestors of the blond aliens who visit Earth today, may have interbred with local humans long ago. Hence, an alien genotype may have entered the human gene-pool thousands of years ago, before the Gaelic people migrated to Ireland, Scotland, northern Spain and Iceland. How could such a major event go unrecorded in human history? If ancient Ireland was invaded by tall blond aliens, who interbred with local humans to leave a hybrid cross, why do historical records not state it clearly?

In fact, the historical records from ancient Ireland do record a whole series of invasions: see *Lebor Gabala Erren* or *The Book of the Taking of Ireland*, compiled in 1150 A.D. The most relevant of these might be an invasion in 1000 B.C. by the mysterious Tuatha de Danaan or Tribe of Danu, who were expert in the arts of pagan cunning. These tall, blond or red-haired strangers supposedly interbred with the locals, while teaching them many useful skills. The *Lebor Gabala* records their dramatic entrance to Ireland as follows:

"In this wise they came, in dark clouds from northern islands of the world. They landed on the mountains of Conmaicne Rein in Connachta, and they brought a darkness over the sun for three days and three nights. Gods were their men of arts, and non-gods their husbandmen."

Evidently the Tuatha de Danaan were advanced enough to arrive in western Ireland near modern Connacht by air; and they were divided into two social classes: "gods" as teachers of medicine, smithing, communication, music or druidry, and "non-gods" as farmers or shepherds. Although no one knows what the Tuatha looked like, we see today depictions of their female leader Eriu as a tall attractive woman with pale white skin, high forehead, long red hair and large slanted blue eyes (see also *Alien Base*, 1997, by Timothy Good). If the Tuatha cross-bred with local humans, they would have left hybrid descendants who look somewhat like themselves. Do not the western Irish and their Gaelic relatives today, often show a high forehead, blond or red hair, and blue eyes? Finally, the mitochondrial DNA as found in our "alien" hair and its related haplotype "hg 1" both cluster around modern Connaught where the Tuatha supposedly landed.

Further evidence as to the nature of the Tuatha de Danaan might be obtained by genetic testing of the tall, fair-haired mummies found at Taklimakan in the western Chinese desert. Those mysterious mummies date to 1200 B.C., and bear European motifs on their arts and clothing. Preliminary analysis of just a few samples from Taklimakan shows a substitution 16,278 C-to-T (Francalacci, P., *The Bronze Age and Early Iron Age Peoples of Eastern Central Asia*, pp. 537–547, 1998), which is precisely the same as that found among blondes of northern Scotland and the Orkney Islands, and also in our "alien" hair root (Table I).

What Do Modern Geneticists Believe?

As an aside, how might modern geneticists in our leading universities view these same data? Firstly, it must be noted with some discouragement, that they are highly dismissive of all UFO-related phenomena, and so are unlikely to take such evidence seriously. Secondly, most of them follow Darwin's model from 1860, where it is assumed that modern humans evolved in isolation on this little planet Earth, from some unidentified lower primate. Thirdly, many of them believe that modern man evolved in Africa or the Near East 200,000 years ago, from whence he spread in the Neolithic Period to other places such as Spain or Ireland.

The present experimental data therefore do not fit within their world view, because they do not believe in UFOs, nor in the almost-human species who seem to inhabit them. [Recall how it was impossible for scientists of the 18th century to believe in meteorites or dinosaurs.]

For example, one professor of human genetics at Oxford has just offered his services over the Internet (http://www.oxfordancestors.com), to trace people's lineages back through time using their Y or mitochondrial DNA. He assumes that all modern humans migrated into Europe from the Near East, and that they descend from just seven women! By that view (news release of April 20, 2000), a female ancestor of the Irish Gaels would look something like the model Carla Bruni: as tall with long blond-brown hair, slanted blue eyes, a high forehead, long thin nose, and prominent chin.

However, not all geneticists agree with him: see Barbujani, G., et al., *American Journal of Human Genetics* 62, 488–491, 1998, or Simoni, L. et al., *AJHG* 66, 262–278, 2000. Nor does his proposed model include historical evidence from the *Lebor Gabala,* that modern Gaels are a cross between the Tuatha de Danaan who came from northern islands, and Celtic Milesians who came from Spain. We would agree that the ancestor of modern Gaels might have looked like Carla Bruni: but did she really walk or sail to Ireland from the Near East, tens of thousands of years ago, after being an African ape further back in her germline? Or more plausibly, was she a cousin of the blond aliens who still visit Earth today?

Advanced Cloning Technology Among the Blond Aliens?

Now somewhat anomalously, our blond "alien" hair also shows for two portions of its hard shaft, a different mitochondrial DNA of Chinese genetics, with substitutions from the consensus at 16,304 (this study), or all of 16,108, 16,129, 16,162, 16,172 and 16,304 (last study). Such a result, even if experimentally confirmed, seems confusing. If that hair shaft really contains Chinese mitochondrial DNA, why is it not black as for all Asian hair, including hair of the second alien reported in 1992?

An important point to note here, is that the color (melanin) of any hair comes from a gene in its nucleus, not from its mitochondrion. Normally the nucleus of any cell matches closely its own mitochondrion, in terms of color and racial type. However, we have just learned on Earth how to perform simple cloning techniques, so that the nucleus of any cell may be transferred to a different cell with a different mitochondrion. For example, clones of Dolly the sheep contain the same nucleus as the original animal, but a different mitochondrion. This is analogous to grafting in plants, where the upper part of some fruiting tree may be joined onto a more durable rootstock, with slightly different genetics.

Our data seem to favor a hypothesis, where the blond alien at some point in her life may have received implants of new hair, by cloning techniques which are just now being developed on Earth as cures for baldness (Hoffman, R., *Nature Biotechnology* 18, 20, 2000). That would explain why her root tissue shows the mitochondrial DNA of a blond, blue-eyed racial type, while her hair shaft (although clear) shows a different kind of mitochondrion. Transfer of blond nuclear DNA to a Chinese hair cell in tissue culture would produce such a result.[205]

Recall from Peter's encounter with the blond alien, that her clear thin hair was very sparse, while the dark hair of her Asian friend resembled a wig. Can we be sure then that either of those women, when visiting Peter in 1992, possessed the hair they were born with? Especially since severe damage to a breast nipple of the blonde, bitten during the encounter, caused no pain; yet it nearly choked the young man, who said it tasted like rubber and stuck in his throat? All of these observations seem consistent with biomedical engineering within the hair and breasts of those alien women.

Is the Blond "Alien" Genetically Resistant to Viruses?

Our studies of the nuclear gene CCR5 seem far more provocative. If authentic, our PCR bands imply that the blond alien might be homozygous for delta-CCR5, and hence genetically resistant to viruses such as HIV or smallpox. That would have profound implications for the nature of our relation with human-like species elsewhere.

The recent origin of delta-CCR5 in man suggests a breeding mutation rather than natural selection, as a possible mechanism of appearance. Thus, one must delete both CCR5 genes at once to achieve a viral-resistant phenotype, which makes a spontaneous origin hard to believe. Could such a double-deletion have been induced by germline genetic engineering, as a way of making a whole alien species viral resistant? Alternatively, if delta CCR5 arose on Earth only recently by random mutation, why did this valuable phenotype not show itself first in apes, or in primitive countries such as India or China where smallpox is rife? How could resistance to HIV have been achieved by natural selection 5000 years ago, if HIV came into man only recently in 1955?

Today we see delta-CCR5 almost exclusively among people of northeast European descent, as well as Ashkenazi Jews (Martinson, J. et al., *Nature Genetics* 16, 100–103, 1997; Libert, F. et al., *Human Molecular Genetics* 7, 399–406, 1998). The origin of that mutation dates to 5000 years ago, and it is associated with a blond, blue-eyed racial type. Could such a mutation have been introduced into local humans, by cross-breeding with some blond alien species in ancient Ireland or Scandinavia?

Unfortunately, sequencing of PCR products from the blond alien hair does not conclusively show whether or not she might be homozygous for delta-CCR5. The sequences amplified within that region are specific for CCR5, yet they do not match the human deletion CATACA-TTAA for both ends within the same clone. An alternative interpretation might be that CCR5 has shortened to two specific sizes, by recombination or jumping during PCR. Only further work on other samples can determine whether the blond alien species might be homozygous for deleted CCR5, and hence genetically resistant to viruses.

Peter Khoury shows two intact genes for CCR5, as does his wife and all members of the investigating team (except for one who is heterozygous); so there is no chance that the blond alien hair root may have been contaminated with local nuclear DNA. Indeed, we cannot find an intact gene for CCR5 anywhere in her genome.

Conclusion

To conclude, this genetic analysis of a reportedly-alien, thin blond hair, has raised a number of important questions for further work. It generally supports the reality of such extraterrestrial contacts, in a Sydney suburb in 1992 and perhaps elsewhere, owing to the rare morphology of such a hair, plus the rare nature of DNA sequences obtained.

Had that hair been a common Earth artefact, it would have shown up early in the investigation. We cannot prove it is an alien hair, yet Occam's Razor supports that notion. Where else in Sydney could Peter Khoury have obtained a long, clear, exceptionally-thin head hair, which shows Basque-Gaelic DNA in its root, but Chinese DNA in two parts of its shaft?

Further work on other potentially-alien biological samples should be undertaken without delay. These studies might identify DNA sequences which are not just rare, but unknown to any plant or animal on Earth. For example, a "yowie" mitochondrial DNA would fit those criteria, if it turned out to contain DNA distinct from that of any local primate. What could be more elegant and simple, than to find DNA as a genetic molecule all through the galaxy?

[1] *Blood Evidence* by Dr. Henry Lee and Frank Tirnady (2003), viii.

[2] Ibid, viii.

[3] Ibid, viii–ix.

[4] While a hair sample is the focus of this book's DNA story, it is intriguing to point out that a coffee cup and hair could have featured in another case, although no DNA work was done to my knowledge. See the bizarre story of "Sally" in Budd Hopkins and Carol Rainey's book *Sight Unseen* (2003) pp. 254–256. Be it fantasy, hallucination, or fact, "Sally" and her experience are described.

[5] Lee and Tirnady, x.

[6] The debate about the reality of alien abductions can be reviewed in a number of worthwhile books. These include *Alien Abductions* by Peter Brookesmith (1998), *The Abduction Enigma* by Kevin Randle, Russ Estes, and William Cone, Ph.D. (1999), Jim Schnabel's *Dark White* (1994), *The Complete Book of Aliens and Abductions* by Jenny Randles (1999), *UFOs and Ufology* by Paul Devereux and Peter Brookesmith (1997), *Alien Discussions* edited by Andrea Pritchard, David

Pritchard, John Mack, Pam Kasey, and Claudia Yapp (1994), *Close Encounters of the Fourth Kind* by C.D.B. Bryan (1995), *Alien Abductions* by Terry Matheson (1998), and *UFOs and Abductions,* edited by David Jacobs (University Press of Kansas, 2000).

[7] Lee and Tirnady, xxiii.

[8] It needs to be made clear that Kary Mullis *did not* carry out the PCR DNA testing of the Khoury hair samples detailed in this book. I made him aware of this work when I interviewed him about his "alien abduction" story.

[9] For an interesting discussion of Locard's principle, see the prologue of Dr. Zakaria Erzinclioglu's book *Every Contact Leaves a Trace.*

[10] For some insight into the real science of forensics reflected in TV shows like *C.S.I.,* see, for example, *The Forensic Science of C.S.I.* by Katherine Ramsland (2001). Popular books on forensic science include *Every Contact Leaves a Trace* by Dr. Zakaria Erzinclioglu (2000), *Hidden Evidence* by David Owen (2000), *Crime Scene* by Richard Platt (2003), *The Encyclopedia of Forensic Science* by Brian Lane (1992).

[11] The *X-Files* has generated a number of books focusing on the "science" behind the show. The best of these is by the science adviser to the show, Dr. Anne Simon, *The Real Science Behind the X-Files* (1999). Another two are *The Science of the X-Files* by Jeanne Cavelos (1998) and the same title by Michael White (1996).

[12] "Nocturnal Visitors" by John O'Neill, *The Skeptic,* Spring 2000, Vol. 20, No 3, pp. 30–33.

[13] *Fire in the Brain* by Ronald Siegel (1993) pp. 83–90; see also *Alien Abductions* by Peter Brookesmith (1998) pp. 125, in the context of a fuller discussion of alien abductions and explanations for them. *The Terror that Comes in the Night* by David Hufford (1982) provides a fascinating cultural perspective on these experiences.

[14] "Memory Distortion in People Reporting Abduction by Aliens," by Susan Clancy, Richard McNally, Daniel Schacter, Mark Lenzenweger, and Roger Pitman, *Journal of Abnormal Psychology,* 2002, Vol. 111, No. 3, 455–461.

[15] Ibid., 456.

[16] See my two-part article "Taken Down Under: The Australian Abduction Experience," *International UFO Reporter (IUR),* Summer and Fall 2003 issues, which refers to some of my early research, in-

cluding my papers "Beyond the CE3 Down Under" (1977), "Australian 'Interrupted Journeys' " (1979) (published in the *MUFON UFO Journal*, No. 150, August 1980, and "Alien Abductions: A Shamanic Perspective of UFOs" in *Nature and Health* magazine, August 1990.

[17] Later in 1992 when Peter Khoury joined the UFO group there was speculation that he had an alien implant. In the wake of the job site assault the group noted in its May 27, 1992, meeting minutes: *"Peter K—recently hit on head with a spade; cat scanned 27/5 (results tomorrow)—may have disturbed possible implant."*

[18] The search for evidence of alleged alien "implants" will be covered later. In another Australian example, during 1990 Keith Basterfield investigated a South Australian woman called "Susan" who revealed "a complex of life-long abductions." Susan also reported that, like a number of others around the world, she may have had an "alien implant," this one placed in her upper mouth. Keith challenged Susan to get a dental X-ray. He didn't expect that anything abnormal would be found. Something rather strange did turn up on the X-ray. Keith spoke with the dentist and eventually obtained the X-ray photo. Keith indicates, "It looked like there was some kind of thin particle in her mouth in the place where she claimed to have an implant. The dentist decided to send Susan to a radiographer to have a full face X-ray taken, but this did not show anything abnormal. Later I had the original X-ray examined by an experienced radiographer, who stated that the X-ray did in fact have something on it but he believed it was merely an artifact of the X-ray process and not indicative of anything actually in the mouth." See *UFOs* by Keith Basterfield (1997), pp. 109–114.

[19] See my book *The Oz Files* (1996) for details on Australian physical evidence cases; Jerome Clark's *The UFO Book* (1997), *The UFO Encyclopedia* 2nd edition (1998), or John Schuessler's *The Cash-Landrum UFO Incident* (1998) for details of the Cash-Landrum case; and Jacques Vallee's *Confrontations* (1990) and Bob Pratt's *UFO Danger Zone* (1996) for information on the "chupa" phenomenon. See also *The UFO Evidence* by Richard Hall (2001), and *The UFO Enigma* by Peter Sturrock (1999).

[20] Budd Hopkins's investigation of the abduction milieu of "Linda Cortile," another alleged multiple-witnessed abduction, described in

his book *Witnessed* (1996), also had its share of frustrations and problems. The reality of the episode, with its staggering implications, if true, is still debated.

²¹ Personal communications.

²² My own account of the complex episode appeared in the *International UFO Reporter*, September/October, 1994, in "An Extraordinary Encounter in the Dandenong Foothills." I also referred to the case in some detail in my 1996 book *The OZ Files*. Kelly Cahill wrote her own book about the experience—*Encounter*—which was published later in 1996. See also *UFOs: A Report on Australian Encounters* by Keith Basterfield (1997), pp. 123–128. Timothy Good provides a good summary in *Unearthly Disclosure* (2001), pp. 51–66. See also my summary "Kelly Cahill Abduction" in *The Encyclopedia of Extraterrestrial Encounters* edited by Ronald Story (2001), pp. 291–293, also published as *The Mammoth Encyclopedia of Extraterrestrial Encounters* in the UK (2002). Phenomena Research Australia (PRA) have yet to publish their account in the broader public UFO literature despite the passage of a decade! Readers are encouraged to write to PRA for further information on Kelly Cahill's case. Perhaps your interest might encourage PRA to more openly share their information. Phenomena Research Australia (PRA) P.O. Box 523, Mulgrave, Victoria, Australia, 3170 or Praufo@aol.com. Along with Peter Khoury's 1992 DNA case, Kelly Cahill's case appears in *The World's Best UFO Cases* compiled by MUFON UFO Journal editor Dwight Connelly (2004). Despite the lack of detailed PRA disclosure and his limited experience with them Connelly describes the PRA group as one of the most respected in the world, a claim I and many other researchers find very difficult to fathom.

²³ Dr. Allen Hynek devoted a chapter of his classic study *The UFO Experience* (1972) to the problem, namely "Science Is Not Always What Scientists Do."

²⁴ Robert Shapiro, *Planetary Dreams* (1999). See also for example *Life Everywhere* by David Darling (2001), *The Extraterrestrial Encyclopedia* by David Darling (2000), and *Life on Other Worlds* (1998) and *The Biological Universe* (1996) both by Steven Dick.

²⁵ Shapiro, 247–248

²⁶ While there are notable positive examples of open scientific inter-

est in the UFO field, these are vastly outnumbered by less helpful and occasionally unfortunate responses and experiences. Ann Druffel's book *Firestorm* (Wildflower Press, 2003) provides a powerful and sobering example of the risks and problems for scientists becoming involved in the UFO subject. Of particular interest and worth in this area are the following books: *The Lure of the Edge* by Brenda Denzler (University of California Press, 2001), *At the Threshold* by Charles Emmons Ph.D. (Wildflower Press, 1997), and *UFOs and Abductions,* edited by David Jacobs (University Press of Kansas, 2000). Jacques Vallee's journals (1957–69) published as a book, *Forbidden Science* (Marlowe, 1996), provide a fascinating insight into the controversy of science and UFOs. Maybe one day UFOs and alien abductions will no longer be a fringe or "forbidden science."

[27] Anomaly Physical Evidence Group (APEG), P.O. Box 42, West Pennant Hills, NSW, 2125, Australia.

[28] Wilson, M.R., et al. (1995) "Extraction, PCR Amplification and Sequencing of Mitochondrial DNA from Human Hair Shafts," *Biotechniques* 18, 662–669.

[29] "Strange Evidence" by Bill Chalker, *International UFO Reporter (IUR)* Spring 1999. See also "UFO Abductions & Science: A Case Study of Strange Evidence" by Bill Chalker, in *Australasian Ufologist,* Vol. 3, No. 3, 3rd Quarter, 1999.

[30] This was not evidence of a DNA phenomenon called "heteroplasmy" (where two different mitochondrial DNAs rarely appear within the sample, usually a result of coexistence of mutant mitochondrial and "wild-type" DNA molecules within a cell or tissue). Heteroplasmy, which is more readily found in human hair than other parts of the body, refers to single-base transitions in the mitochondrial DNA. For example G to A, or one C to T. They are not big changes. Environmental exposure and aging can be factors.

[31] Hoffman, R., "The Hair Follicle as a Gene Therapy Target," *Nature Biotechnology,* Vol. 18, January 2000, 20–21.

[32] Martinson, J., et al., *Nature Genetics* 16, 100–103, 1997; Libert, F., et al., *Human Molecular Genetics* 7, 399–406, 1998.

[33] See *Alien Identities* by Richard Thompson (1993), pp. 437–446.

[34] "The Unusual Origin of the Polymerase Chain Reaction," by Kary Mullis, *Scientific American,* April 1990. The complex evolution of

the PCR process and Kary Mullis's role is covered in the book *Making PCR* by Paul Rabinow (University of Chicago Press, 1997). A briefer account in the context of the extraordinary contribution of DNA analysis is described in "The Improbable Origins of PCR," a chapter in *Blood Evidence* by Dr. Henry Lee and Frank Tirnady (Perseus Publishing, 2003).

[35] This is the extent of the story Kary Mullis told in his book *Dancing Naked in the Mind Field* (Bloomsbury, 1998), Chapter 13 "No Aliens Allowed." The rest of his story came out in my interview with him on April 4, 1999.

[36] See Michio Kaku, *Hyperspace* (Oxford University Press, 1994).

[37] See David Jacobs, "The Threat" (Simon & Schuster, 1998).

[38] Two other members of UFO Research (NSW)—Frank Sinclair and Coralee Vickery—undertook an earlier preliminary interview with Peter Khoury on March 7 1992. It was a very tentative interview and Frank Sinclair advised me that it was his first significant interview. The UFO sighting with Vivian is discussed along with some discussion of the 1988 episode. Nothing else was raised. Frank Sinclair kindly made a copy of the tape available to me during 2003. It confirms that the details described are consistent with Peter Khoury's later recollections about the UFO sighting and the 1988 abduction episode. There was some confusion about the date of the abduction episode, whether it was 1988 or 1989, but the interview process did not extend to an examination of this aspect. It is clear from all the information we have from diverse sources that Peter's experience with the needle probe occurred in 1988. There was nothing in this basic interview about any abduction style experience between 1988 and March 1992 other than the 1988 episode we know about.

[39] This is the sample reported in the second phase of DNA work re the Khoury hair sample as a "putative yowie sample," which was utilised as a secondary "control" sample. See Appendix.

[40] John Mack authored *Abduction* (1994) and later *Passport to the Cosmos* (1999).

[41] It is interesting to see Peter's description of a feeling of a presence in his house for some time after the explicit encounter. Was this simply a predictable psychological response or something else? In an interesting study John Auchettl of PRA described their controlled

monitoring of events at "Tyers farm" in the Moe district of Victoria, Australia, utilizing temperature monitoring and a near-field acoustic holography (NAH) system to substantiate the presence of "invisible boundary" or "energy flow" effects. See "The Tyers Farm Event—Invisible Boundary Window" by John Auchettl, *UFO Quest*, Issue #1, October 2003, pp. 4–11, 19.

[42] Incidents in which people claim to have been transported through solid material, like walls, glass windows, and mirrors have occurred with some frequency. Even Peter Khoury's detail of the beings coming through the mirror has been reported. For example, Rusty Hudson, a young band member staying in a hotel in Florida during 1980, reported that a being with large slanted black eyes emerged from an intense burst of light that came from the mirror in his room. He apparently was taken somewhere and examined. He next found himself outside the hotel room door. His fellow band members were totally confused as they had witnessed the light burst and verified that he had suddenly gone missing under very strange circumstances. See the *Reader's Digest* publication *UFO: The Continuing Enigma* (1991) pp. 9–17. There is a long and continuing tradition of the utilization of mirrors as "gateways" or "technology" for manipulating states of mind or entering other realities. See Mark Pendergrast's book *Mirror|Mirror* (2003). For more recent visionary takes utilizing mirrors see also *Reunions* (1994) by Raymond Moody with Paul Terry, and *Wisdom from the Great Beyond* by Jan von Denmarc, Ph.D. (2003).

[43] The detail and the debate about "aliens" can be followed in books such as *The Field Guide to Extraterrestrials* by Patrick Huyghe (1996), *Aliens* by Jenny Randles (1993), *Faces of the Visitors* by Kevin Randle and Russ Estes (1997), and *The Complete Book of Aliens and Abductions* by Jenny Randles (1999). Jerome Clark provides a good overview of the topic in the "Close Encounters of the Third Kind" entry in his excellent *UFO Encyclopedia* (1998) (see Vol. 1, pp. 207–239). Further reviews of this area can be found in Richard Hall's excellent book *The UFO Evidence, Volume II* (2000), and in some of the "alien" entries in *The Encyclopedia of Extraterrestrial Encounters* edited by Ronald Story (2001). The classic studies in this controversial area are the following books: *The Humanoids* Edited by Charles Bowen (1969)

and *Flying Saucer Occupants* by Coral and Jim Lorenzen (1967) (reissued in a revised and expanded format as *Encounters with UFO Occupants* in 1976).

[44] Many readers would have seen the Wandjina rising from the ground in an enormous and spectacular display in the opening ceremony of the Sydney Olympic Games in 2000. See *The Art of the Wandjina* by I. M. Crawford (1968), *Expeditions in Western Australia, 1837–1839* by George Grey, *Yorro Yorro* by David Mowaljarlai and Dutta Malnic (1993).

[45] See Whitley Strieber's *Transformation* (1988), pp. 121–122. Strieber in quoting the woman leaves out any reference to the hair she observed.

[46] See *Breakthrough* (1995) by Whitley Strieber, pp. 110–112. Seven years later Strieber refers to the beings in this case having hair. He quotes from the woman's letter from 1987–88: "Unlike your visitors, mine had wispy bits at the back of their heads" (p. 111). The hair on the two "males" was brown and the female "black," with color-coordinated "shrouds." Strieber doesn't link the account here to the reference in his previous book, nor does he highlight that he had kept the presence of hair out of his own original "visitor" descriptions.

[47] See *The Communion Letters* by Whitley Strieber and Anne Strieber, (1997), pp. 53-55. Even though Strieber has the woman's letter in the chapter entitled "Alien Abduction as It Really Is" and that it is "the most vivid description of the visitors in all our records" he did not reveal that his own original "visitors" featured hair.

[48] See *Abduction* (1988) and *MIB* (1997) by Jenny Randles.

[49] See Eddie Bullard's excellent essay on abductions in Jerome Clark's monumental work *The UFO Encyclopedia* (1998): "Abduction Phenomenon" by Thomas E. Bullard, pp. 1–26. Bullard focuses on the "first" cases on p. 2.

[50] See John G. Fuller's book *The Interrupted Journey* (1966). "Hill Abduction case" in Jerome Clark's *UFO Encyclopedia* (1998), Vol. 1, pp. 489–504 gives a good summary of the case.

[51] Fuller, p. 296.

[52] For the earliest accessible accounts of Villas-Boas' experience see "Report on the Villas-Boas Incident by Olavo Fontes, M.D., & João Martins, translated by Irene Granchi" in *Flying Saucer Occupants* by Jim

& Coral Lorenzen (1967) and "The Amazing Case of Antonio Villas Boas" by Gordon Creighton in *The Humanoids* edited by Charles Bowen (1969). "Villas-Boas CE3" in Jerome Clark's *UFO Encyclopedia* (1998), Vol. 2, pp. 974–977 gives a good summary of the case.

[53] "The Melancortin 1 Receptor (MC1R): More Than Just Red Hair" by J. L. Rees, *Pigment Cell Research,* 13: 135–140 (2000); "Human Pigmentation Genetics: The Difference Is Only Skin Deep" by R.A. Sturm, N.F. Box & M. Ramsay, *Bioessays,* 20: 712–721 (1998); "What Controls Variation in Human Skin Color?" by G. Barsh, *PloS Biology,* 1/1:19–22 (2003).

[54] See "Ready for Your Close-up?" by Clare Wilson, *New Scientist,* July 20, 2002, pp. 34–37. The article reports on the partial success of biotech firm DNAPrint Genomics of Sarasota, Florida, mainly in eye color so far. DNAPrint are also applying their techniques, which utilize single nucleotide polymorphisms (SNP) and candidate genes, and are already tentatively claiming they can classify people as blond, red, auburn, brown, or black-haired from their DNA alone. As researchers unravel this complex area, it may provide some interesting genetic markers to follow in our area of controversy. See also the fascinating and informative work by Armand Marie Leroi: *Mutants* (2003).

[55] FSR 37/1 Spring 1992

[56] "The New A.V.B. Case from Brazil: Full Account" by Luiz do Rosario Real, *FSR* 31/3, 1986, pp. 16–24, also reported in *FSR* 30/5, 1985.

[57] See "Abduction at Botucatu" by Irene Granchi, *FSR* 30/1, 1984, pp. 22–25.

[58] *Sexual Encounters with Extraterrestrials* by C.L. Turnage (2001) pp. 73–85.

[59] Ibid. pp. 59–72.

[60] *The Love Bite* by Eve Longren (2000).

[61] See, for example, some of the dynamics played out in *Witnessed* by Budd Hopkins (1996) and perhaps in the Andreasson-Luca relationship (both were abductees) described in the Andreasson/"Watcher" series of books by Ray Fowler.

[62] Michael Swords has presented some excellent reviews of this problem in "Extraterrestrial Hybridization Unlikely" (*MUFON Jour-*

nal, November, 1988) and "Modern Biology and the Extraterrestrial Hypothesis" (*MUFON 1991 International UFO Symposium Proceedings,* abbreviated in "Modern Biology and Extraterrestrials," in Bill Fawcett's *Making Contact,* 1997). An entertaining discussion of the problems can also be found in Jeanne Cavelos' book *The Science of the X-Files* (see "Grays, Hybrids, and UFOs").

[63] Transgenic research and its implications have been, and are being, extensively addressed. While the Internet is a good source for up-to-date developments, the following sources provide a variety of coverage from the general, specific, or conjectural: *GE—Genetic Engineering and You* by Moyra Bremner (1999), *Remaking Eden* by Lee M. Silver (1998), "Dissecting and Manipulating Genes" chapter in *Human Molecular Genetics* by Tom Strachan and Andrew Read (1997), "Playing God—Designer Children and Clones" chapter in *Visions* by Michio Kaku (1998).

[64] *Sight Unseen* by Budd Hopkins and Carol Rainey (2003), p. 254.

[65] Ibid. p. 256.

[66] Ibid. p. 257.

[67] Ibid. p. 255.

[68] Ibid. pp. 255–256. Jim Schnabel, in his book *Dark White* describes a more complex tableau surrounding Sally, whom he calls "Lucy." While his version more strongly suggests the specter of delusion, Schnabel does write, "Although eerily ageless, he was flesh and blood. She once bade him remove his shirt, and plucked blond hairs from his chest." Schnabel's version of "Stewart" (whom he calls "Steven") seems almost totally preoccupied with sexually abusing "Sally" ("Lucy") often to the mantra, "You will enjoy this," p. 250. The abductee "Deborah" in David Jacob's book *The Threat* (1998) pp. 186, 198–202, may also be "Sally." She experiences a strange "job interview" and has lifelong interactions with an "alien sadist." The picture Schnabel paints of "Lucy" (aka "Sally") does not sit well with the individual Budd Hopkins described to me. Hopkins told me that he felt the subject frightened Schnabel at times, with the end result being that Schnabel turned on it, debunking the whole subject.

[69] The Australian UFO Research Network, PO Box 738, Beaudesert, Qld 4285, Australia, http://www.hypermax.net.au/~auforn. The toll-free number, which only operates in Australia, is: 1800

772288. In the US contact the Center for UFO Studies (CUFOS): www.cufos.org or the Mutual UFO Network (MUFON): http://www. mufon.com. For worldwide resources try John Hayes' excellent site: http://www.ufoinfo.com/contents.shtml. *The Little Giant Encyclopedia of UFOs* by Jenny Randles (2000) is a good overview of the UFO subject and ufology.

[70] UFOR (NSW) assisted with the funding of these preliminary PCR "human consensus" DNA comparisons.

[71] Harry Turner's role is described in my document "UFOs Sub Rosa Down Under" (1996) which can be accessed on my web site: www.theozfiles.com. See also my book *The OZ Files* (1996).

[72] CSIRO—Commonwealth Scientific and Industrial Research Organisation—Australia's main government science and industry research organization.

[73] See "The Cressy Affair" in *The Oz Files* (1996), pp. 103–106.

[74] This was probably the highly publicized sighting by air traffic control personnel at Canberra airport on July 15, 1965. There were attempts to link this sighting to problems that nearby Tidbinbilla Tracking Station was having with the Mariner 4 spacecraft flyby of Mars. Both the reported problems with communications with Mariner and the UFO sighting were probably spurious. The evidence for an astronomical explanation was strong.

[75] The description of an entity with "a head more like a dog" could be like the entity described in two striking encounters in Argentina in September 1972, in a large motor manufacturing plant. Both witnesses described an entity with a flat nose and pointy ears. See "The Anthropomorphic Phenomena at Santa Isabel" by Oscar Galindez, Parts 1 & 2, *FSR*, Volume 21, No. 2 1975 and Volume 21 No. 3 & 4 (Double issue) 1975.

[76] See "Apocalyptic Thought Connected with UFOs and ETs" by Martin Kottmeyer in *The Encyclopedia of Extraterrestrial Encounters* (2001) edited by Ronald Story, pp. 77–83; an evocative example of this type of milieu can be seen in *The Mothman Prophecies* by John Keel (1975). See also *Expecting Armageddon* (2000) edited by Jon Stone, *Apocalypses* (1999) by Eugen Weber, *Countdown to Apocalypse* (1998) by Paul Halpern, *Destroying the World to Save It* (1999) by Robert Jay Lifton, *End-Time visions* (1998) by Richard Abanes, *Living at the End of*

the World (1999) by Marina Benjamin and *Children of Ezekiel* (1998) by Michael Lieb.

[77] Vicki had another child, a daughter.

[78] Terrible fires did decimate suburbs of Canberra on January 18, 2003, even severely damaging the historic Mt. Stromlo observatory, site of a 1957 astronomer UFO sighting. See my book *The OZ Files* (1996) pp. 86–87.

[79] This connection is covered in the following works: "When the Snakes Awake: Animals and Earthquake Prediction" by Helmut Tributsch (1982). See also the work of Paul Devereux, in particular, "Earth Lights: Towards an Understanding of the UFO Enigma" (1982) and "Earth Lights Revelation: UFOs and Mystery Lightform Phenomena: The Earth's Secret Energy Force" (1989). See also "UFOs and Ufology—The first 50 years" by Paul Devereux and Peter Brooke-smith (1997), particularly Chapter 10: "Planet Earth's UFOs—Chasing the Wild Lights."

[80] UFOIC files.

[81] An article in the Sydney newspaper *The Sun-Herald*, "Earthquake Alert" by Leon Oberg, August 20, 1989, elaborates on Vicki's earthquake system: "Vicki Klein believes animals have a built-in earthquake alert system following an experience in 1977 at her home in Dalton. . . . On that occasion, all four family owned horses acted abnormally for several hours before a strong earthquake hit the district. 'They all stood silently facing north-east and despite being turned around they immediately swung back to the north-east in a compass-like action,' she said. 'Two hours later an earthquake measuring five points on the Richter scale hit.' . . . Mrs. Klein devised a homemade earthquake measuring scale during the 1970s which was based upon the sounds of earthquake vibrations. Mrs. Klein, a keen musician, described musical pitch as vibration. She sensed the varying pitches (vibrations) during successive shakes and devised her own points system. Following the experience with her horses during the 1977 quake, Mrs. Klein started to take a 'keener, more scientific' interest in the rumblings from the ground and responded to a circular hanging in her local post office from the Bureau of Mineral Resources which was seeking people to take recordings. 'When the Bureau learned I had devised my own scale, they were extremely interested and after

they compared notes with the official measurements, we were both stunned to find how very close my readings were to theirs,' she said."

[82] The "star children" belief has emerged from the "star people" concept popularized by writers such as Brad Steiger. (See *The Star People*, 1981 and *The Seed*, 1983. Steiger dwelt on aspects of this milieu in his 1976 book *Gods of Aquarius*, specifically chapter 7 "The Star Maidens and the worldwide production of little Uri Gellers.") Dr. Richard Boylan has emerged as a controversial advocate of "star children" (see *The Abduction Enigma* by Kevin Randle, et al., for some background on the controversy surrounding Richard Boylan). Jenny Randles provides intriguing material in her 1994 book *Star Children*. Whitley Strieber touches on this theme in some of his books, in particular *The Secret School* (1997). *Awakening* (2002) by Mary Rodwell provides a somewhat uncritical and New Age "bible" for experiencers. See in particular her Chapter 9 "Star Children—'Homo Noeticus,' The New Humans, or 'New Kids on the Block.'" Colin Wilson in *Alien Dawn* (1998) concludes with a similar focus. Some of the directions this belief system are leading to are described in such books as *From Elsewhere* by Scott Mandelker, Ph.D. (1996) and *Aliens Among Us* by Ruth Montgomery (1985). An interesting journey through some of this territory can be found in *Soul Samples* by Leo Sprinkle, Ph.D. (1999). Researchers and travelers in this controversial area of "star children" claims would do well to take on board the lessons of such books as *Hystories* (1997) by Elaine Showalter and *Sleeping with Extra-terrestrials* (1999) by Wendy Kaminer. *In Advance of the Landing* by Douglas Curran (1985) provides a poignant "road trip" on this highway to the "aliens-r-us" or the "alienated."

[83] Dr. Stephen Larsen also co-authored the authorized biography of legendary mythologist Joseph Campbell, *A Fire in the Mind*, and authored *The Mythic Imagination*.

[84] See also Bradford Keeney's *Shaking Out the Spirits*.

[85] From cover of *Vusamazulu Credo Mutwa* edited by Dr. Bradford Keeney.

[86] Quoted with permission of Rick Martin, from his September 30, 1999, interview with Credo Mutwa.

[87] See *The Song of the Stars* pp. 134–138.

[88] In 1999 David Icke undertook a controversial video interview

with Credo Mutwa (just before the interview with Rick Martin I have quoted from), distributing it as "The Reptilian Agenda," to support his bizarre theories. It is easy to view Icke's "reptilian rants" as extreme paranoid fodder, but Credo Mutwa's accounts seem anchored in the atavistic legacy of African magical beliefs, a potent force in many African cultures. His story and "otherworldly" claims are not just artifacts of Icke's "reptilian" agenda. They have a long lineage of which Icke's involvement is only a recent manifestation. The extent and variety of Credo Mutwa's alien tales seem rather troubling for the credibility of his claims. In particular Icke's appropriation of the "reptilian" tales tests many UFO researchers' patience. Most seem to write Credo Mutwa's tales off as story telling aberrations or just plain fabrication. While Icke's extrapolations boggle the mind or predictably close it, Credo Mutwa's tales may well be something else.

[89] See *Passport to the Cosmos* by John Mack, pp. 196–199, and *Song of the Stars* edited by Stephen Larsen, pp. 141–142.

[90] See *Fire in the Sky* by Travis Walton (1996).

[91] See my book *The Oz Files* (1996) pp. 21–22.

[92] Jenny Randles provides an interesting review of similar phenomena in *Time Storms* (2001).

[93] The late Cynthia Hind, author of *UFOs: African Encounters* (1982) and *UFOs over Africa* (1996), wrote about these Zimbabwean government "disappearances" on Mt. Inyangani in an article for Fate magazine during the early 1980s. My source is an online article by Scott Corrales, "In the Wink of an Eye: Mysterious Disappearances" at www.think-aboutit.com/Misc/in_the_wink_of_an_eye.htm.

[94] Just as Credo Mutwa's tales evoke shamanic perspectives, aspects resonate powerfully and sometimes uneasily with the contemporary UFO vision. Another intersection that has dragged in the alien theme is the controversial field of remote viewing (RV)—or "psychic spying," to put it simply. Recollect that Credo Mutwa's horrific circa 1959 alien abduction experience allegedly took place in the region of Mt. Inyangani in eastern Zimbabwe. Skip Atwater, who was the operations officer for the U.S. Army's Intelligence and Security Command remote viewing project STARGATE from 1978 to 1987, describes an unlikely resonance with Credo Mutwa's alien visions at Mt. Inyangani. In 1973 Pat Price, one of the "stars" of the Stanford

Research Institute's (SRI) genesis RV program, gave an unsolicited package to Hal Puthoff, one of the key scientists running the SRI project. It contained Price's remote viewing notes of four "UFO bases" on Earth. Puthoff didn't know what to think, but he was sure of Price's remarkable psychic skills. (For details on Pat Price see *Mind-Reach* by Russell Targ and Harold Puthoff [1977] and *Remote Viewers* by Jim Schnabel [1997]). One of the bases was alleged to be located just to the west of the secret U.S. spy base in central Australia—Pine Gap. Because he knew the then CIA station chief of Pine Gap, Puthoff contacted him, asking if anything unusual was going on in the area. His friend said there wasn't much going on beyond the routine of intelligence gathering, except for maybe the UFO activity that was being reported around Mt. Zeil, which was about 140 kilometres to the west. Puthoff was intrigued, as Price's RV location of the alleged Australian "alien base" was Mt. Zeil! By the early 1980s these notes made their way into the hands of army officer Atwater. Charged with the responsibility of training and assessing the army remote viewers he utilized Price's "targets" in a blind controlled fashion. He was struck when his own army remote viewers supplied him with information that suggested there were hidden underground facilities with some sort of alien or extraterrestrial connection at the four localities Price had viewed a decade earlier. Apart from Mt. Zeil in Australia, the other three "alien bases" were located at Mt. Hayes in Alaska, Mt. Perdido in the Spanish Pyrenees, and Mt. Inyangani in Zimbabwe! See the CD with *Captain of My Ship, Master of My Soul* by F. Holmes Atwater (2001). Other remote viewing players have emerged with alien tales. RV pioneer Ingo Swann came out with an exotic UFO-related memoir played out between 1975 and 1977 in his book *Penetration* (1998). It told of covert meetings with a shadowy group of individuals that led to Swann participating in remote viewing evidence of intelligence on the moon, an L.A. supermarket encounter with a buxom ET female, and secretly observing a UFO over an Alaskan lake. Schnabel's *Remote Viewers* mentions other RV ET interludes. Courtney Brown describes his controversial RV research in two books—*Cosmic Voyage* (1996) and *Cosmic Explorers* (2000). The "ET" dimensions of remote viewing are big on controversial detail and short on credible evidence.

[95] Hardly a practice I would recommend. See *Cannibals, Cows and the CJD Catastrophe* by Jennifer Cooke (1998).

[96] See Timothy Good's book *Alien Base* (1998) Chapter 15, "The Plantation."

[97] See "The Continuing Story of the Imjarvi Skiers" by Anders Liljegren, *FSR*, Vol. 26, Nos. 3 & 5, 1980–81 and "A Nordic Saga" in "The Complete Book of Aliens & Abductions" by Jenny Randles, p. 81.

[98] "The Anthropomorphic Entity at Villa Carlos Paz" by Dr. Oscar Galindez, *FSR*, Vol. 26, Nos. 5 & 6, and Vol. 27, No. 1. Quote from *FSR*, Vol. 26, No. 5, p. 14.

[99] I have seen this mechanism at work in other people, such as Australian aboriginal "abductee" Lorraine Maffi Williams. She has been described as a storyteller and "liaison officer" for the spiritual truths of her Githebul group from the Bundjalung tribe. Lorraine revealed to me some of her perceptions of these intrusions into both aboriginal and western lives:

"Our attitude to what goes on up in the heavens is what rules us Aboriginals. Its similar to religion, whereby Christians believe in a religious world ruled by one God, but many "Saints," we believe the same only the many "saints" to us are planetary ones whom you in the the Western world refer to as ET or aliens. We call them Wandjinas and Mimi Spirits, and have done so for thousands of years, until 1788, when an English concept of the above was interestingly enough (found to be) *parallel to what we have practised for eons; that we commonly refer to as our Dreamtime, that began in the Milky Way."*

Lorraine suggests that communications from these beings is ongoing and cites her own experiences with her "old friend" since the age of 12. She feels she is an "abductee." *". . . My dear old friend took me up, yes in a UFO, but a different sort to the western beliefs."* Her experiences were more spiritually oriented. She adds, *"I went through all or nearly what abductees did. . . ."* Lorraine indicated, *"We believe in UFO, but here to we have the aboriginal concept and belief, and we know about abductions and why."*

One wonders what these expressions on what seems to be interactions with another culture—a native spiritual or an alien culture—might lead to. Such people and their experiences should not be reduced in stature by uninformed criticisms coming from people who are ignorant of the odysseys such people have been on. Nor should there be misguided, blinkered embraces anchored in syco-phantic hagiography. Neither are without their flaws and problems, and neither would the high-minded critics be, if they had had the same "life experiences" of people like Credo Mutwa and Lorraine Maffi Williams to deal with. As people like Credo and Lorraine go through their "journey," of course they will "filter" the ideas of others and maybe their worldview will be diluted somewhat from the origi-nal and unique indigenous worldviews they have presented. These people have tried to create "bridges" between cultures and therefore it is inevitable they may reap the flak of both their own cultures and those "superior" cultures looking in. Lorraine passed away recently.

[100] Elizabeth Klarer claimed contact and liaisons with benevolent "space brother types." Their liaisons grew so intimate that Klarer al-leged she had an ET child, who lived with his father on the alien home planet. See "UFOs—African Encounters" (1982) and "UFOs over Africa" (1997) by Cynthia Hind.

[101] *UFOs: African Encounters* (1982) and *UFOs over Africa* (1997).

[102] The attractive boxed book *Vusamazulu Credo Mutwa* from Ring-ing Rocks Press and edited by Bradford Keeney (2001) is a striking testament to part of that legacy, with a rich presentation of his art and life, along with a CD that includes tributes in song, and his own account of his initiation as a *sangoma*.

[103] See for example the classic study by Mircea Eliade *Shamanism*, Carmen Blacker's *The Catalpa Bow*, Michael Harner's *The Way of the Shaman*, Joan Halifax's *Shamanic Voices*, Holger Kalweit's *Dreamtime and Inner Space*, Piers Vitebsky's *The Shaman*, Shirley Nicholson's com-pilation *Shamanism*, Roger Walsh's *The Spirit of Shamanism*, Gary Doore's compilation *Shaman's Path*, Nevill Drury's *Shamanism*, and Jeremy Narby and Francis Huxley's compilation *Shamans through Time*.

[104] *Dreamtime and Inner Space* by Holger Kalweit (1988), pg. 132.

[105] See for example *Earthmind* by Paul Devereux, John Steele, and David Kubrin (1989), and *Alien Energy* by Andrew Collins (2003).

[106] See for example Paul Devereux's *Places of Power* (1990).

[107] See for example Michael Harner's book *The Way of the Shaman* (1980), pp. 1–8, Jeremy Narby's *The Cosmic Serpent* (1999), *Ayahuasca Visions* (1991) by Luis Eduardo Luna and Pablo Amaringo, "Experi-ences of Light and Balls of Fire" in Holger Kalweit's *Dreamtime and*

Inner Space (1988), and *True Hallucinations* by Terence McKenna (1993).

[108] See for example *Electric UFOs* by Albert Budden (1998)

[109] See for example the work of Paul Devereux, in particular, *Earth Lights* (1982) and *Earth Lights Revelation* (1989). See also *UFOs and Ufology* by Paul Devereux and Peter Brookesmith (1997), particularly Chapter 10: "Planet Earth's UFOs—Chasing the Wild Lights."

[110] *Voices of the First Day* by Robert Lawlor (1991)

[111] Interview with Robert Lawlor, January 1996.

[112] *Vusamazulu Credo Mutwa* edited by Bradford Keeney (2001), pp. 82–83.

[113] *Crime Scene* by Richard Platt (2003), *The Forensic Science of C.S.I.* by Katherine Ramsland (2001), *Every Contact Leaves a Trace* by Dr. Zakaria Erzinclioglu (2000), *Hidden Evidence* by David Owen (2000), and *The Encyclopedia of Forensic Science* by Brian Lane (1992) are accessible starting points for researchers to examine the range of tools and techniques available that may have applications in alien abduction events.

[114] Victoria Alexander quoted Linda Howe from "The UFO Jigsaw" by, *International Symposium on UFO Research,* 1992. See also "My Close Encounters of the Third Kind" by David Huggins, *Leading Edge International Research Group Journal,* Issue No. 46, August–Sept. 1992.

[115] "New Protocol for abduction research" by Victoria Alexander, *MUFON UFO Journal,* No. 307, November 1993, pp. 7–10.

[116] Perhaps an oddity in the UFO literature ranks, *Reveal Nothing* by Kerry Wisner and Linda Rakos (2001) appears to have slipped under the American UFO research radar. The book was published by Hwt-Hrw Publications, which mainly publishes books on Egyptian paganism and magick. Some of these were written by Kerry Wisner. These things may account for the book's low profile. At only 117 pages it purports to feature a forensic hypnosis based investigation of an abduction experienced by one John Williams that allegedly first yielded hypnotic testimony back in 1974. The authors came upon this early material in 1990. Its 1974 content purports to reveal advanced knowledge only now being verified in such areas as genetic engineering, cloning, dark matter, astrophysics, pole shifts, and the origins of the human race. The authors claim corroboration from independent

testimony from a second witness, polygraph reports, doctor and therapist testimony, scientific journals, and recent scientific discoveries. Whether this material is legitimate depends on how well the book and its claims stand up to independent investigations.

[117] See "Anatomy of a Mass Abduction" by Dale Musser, *UFO Universe magazine,* Fall 1993, pp. 24–26, and "The Alien Hunter—Derrel Sims," in *The Abduction Enigma* by Kevin Randle, Russ Estes, and William Cone, Ph.D. (1999), pp. 248–259.

[118] Cheryl Powell of ICAAR was interviewed by Mike Corbin on Paranet Continuum on April 2, 1995, and June 2, 1996. The first interview mentioned the forensic approach and some promising leads but the 1996 interview reported that nothing of note had been found.

[119] See Filer's references to the "crime scene perspective" in his on-line Filer's Files during 1998 and 1999.

[120] See "Betty Hill Dress Analysis," *Technical Service Response* No. UT025, October 19, 2003, and personal communications.

[121] For some details of Dr. Levengood's "crop circle research," see *Mysterious Lights and Crop Circles* by Linda Moulton Howe (2000), *UFOs and Ufology* by Paul Devereux and Peter Brookesmith (1997), and *Crop Circles* by Colin Andrews with Stephen Spignesi (2003).

[122] "Physical Evidence of Alien Abductions May Exist in Household Dust," February 16, 2000, www.abduct.com/aaer2/r9.htm. The ad can be sighted at www.abduct.com/emporium/ad5.htm.

[123] See "DNA testing inconclusive" by Barry Taylor, *MUFON UFO Journal,* November 2000, pp. 6–7, UFO Update at www.virtuallystrange.net/ufo/updates/ for September and October 2000.

[124] *The Aliens and the Scalpel* by Roger Leir (1998), *Casebook: Alien Implants* by Roger Leir (2000), and *Confirmation* by Whitley Strieber (1998).

[125] "Physical Evidence and Abductions," by David Pritchard, *Alien Discussions* (1994).

[126] Ibid. See also "Alien Implant" in *Swamp Gas Times* by Patrick Huyghe (2001).

[127] "The Implant Motif in UFO Abduction Literature" by Keith Basterfield, *Journal of UFO Studies,* New Series Vol. 8, 2003, 49–84.

[128] "Blinded by the Light," *MUFON Journal* No. 373, May 1999, pp. 3–9. The report was reprised in the *2000 MUFON Symposium Proceed-*

ings. See also "Analysis Supports Abduction on Video" by John Carpenter, *MUFON Journal* No. 390, October 2000.

[129] See Ann Druffel, *How to Defend Yourself Against Alien Abduction,* Three Rivers Press, NY, 1998, pp. 79–87 and pp. 196–199.

[130] "Invisibility and the UFO Abduction Phenomenon" by Budd Hopkins, 1993 *MUFON Symposium Proceedings.* See also *Sight Unseen* by Budd Hopkins and Carol Rainey (2003), Chapter 4.

[131] From www.ufoitalia.net plus personal communications.

[132] "Monitoring Abductions" by Mark Rodeghier, *MUFON Journal,* No. 405, January 2002.

[133] Communications from Tom Deuley, June 2003.

[134] Communications from Tom Deuley and Mark Rodeghier, September 2004.

[135] Gary Lowrey's experiences have been investigated by a number of parties, including Dr. Roger Leir. See http://ufoxfiles.com/glowery/update, *Alien Abduction: The Gary Lowrey e-book, UFO magazine* (UK) May 2002, pp. 13–15, and personal communications.

[136] The full report and data can be accessed on the NIDS website: www.nidsci.org.

[137] Dr. Roger Leir provides an overview of the context of the "claw" and the case in "Alien Abduction and Academic Science," *UFO Magazine,* January 2004, pgs. 46–48.

[138] For my own survey of the worldwide UFO physical trace phenomenon, see "Physical Traces" in *UFOs 1947–1987* edited by Hilary Evans with John Spencer (1987), pgs. 182–196. *The UFO Enigma* edited by Peter Sturrock (1999) provides a good review of the physical dimensions of the UFO phenomenon.

[139] See for example "Light Phenomena" in Richard Hall's *The UFO Evidence, Volume II* (2001), pp. 378–389.

[140] See "The Strange Case of Dr. 'X' " (2 parts) by Aime Michel, in *FSR* "UFO Percipients" Special Issue No.3, September 1969, and *FSR* Vol. 17, No. 6, 1971. See also the chapter "Clinical Data" in *Confrontations* by Jacques Vallee (1990). See "Rejuvenation follows close encounter with UFO" by Pedro Romaniuk, *FSR,* Vol. 19, No. 4, July-August 1973, pp. 10–13, 14; "The Extraordinary case of rejuvenation" by Pedro Romaniuk, *FSR,* Vol. 19, No. 5, September–October 1973, pp. 14–15. For further examples see "UFO Healings: True Ac-

counts of People Healed by Extraterrestrials" by Preston Dennett (1996). John Mack reports a rapid healing of a hand injury in the wake of a 1976 encounter between the mantindane and Credo Mutwa. See pp. 199–200, of "Passport to the Cosmos" by John Mack (1999).

[141] Our progress in understanding light and manipulating it is covered in the following references: *The Story of Light* by Ben Bova (2001) and *Light Years* by Brian Clegg (2001). Clegg's book describes the intriguing use of "Bose-Einstein condensates to drag the speed of light back to a crawl." I drew attention to the *Nature* paper that described this striking advance on UFO Update in a message entitled: "'Solid light' UFO Events—A Theoretical Basis?" on January 31 2001: http://www.virtuallystrange.net/ufo/updates/2001/jan/m31-014.shtml. The *Nature* paper: "Observation of coherent optical information storage in an atomic medium using halted light pulses," by Hauu et.al, 25th January, 2001, is available on the *Nature* web site in text and PDF format: http://www.nature.com/nature/fow/010125.html.

[142] "The Gundiah Mackay Abduction Milieu" by Bill Chalker and Diane Harrison, *Australasian Ufologist*, Vol. 5 No. 4 2001, pp. 26–28.

[143] "The UFO Report—I Was Abducted by Aliens" by "Kate Johns," *New Idea*, November 17, 2001, pp. 32–35.

[144] The *"Fraser Coast Chronicle"* series by Kevin Corcoran was published on October 11–12, 2002, with the following headlines: "Why did they do it?" "And where are they now? Beam of light carries woman through window to 1.8 m alien" "If a hoax—police say it was—what reason?" "Mayor reckons affair certainly was no hoax" and "Researchers find story did not add up."

[145] Years before I had read Christopher Evans book *Cults of Unreason* (1973).

[146] See *Bare Faced Messiah* by Russell Miller (1987), 372.

[147] *Wild West China* (2003) by Christian Tyler provides a fascinating journey through the complexities of Xinjiang. The ferment and diffusion effects of the region are evocatively captured in *The Silk Road* by Frances Wood (2003).

[148] Haplotypes are particular combinations of alleles (alternative forms of genes) or sequence variations that are likely to be inherited together. The alleles are closely linked on the same chromosome

(from "Genome speak" in *The Human Genome* edited by Carina Dennis and Richard Gallagher [2001]).

[149]   Race determinations via DNA are even more complex than previously thought. See "Does Race Exist" by Michael Bamshad and Steve Olson, *Scientific American,* December 2003, pp. 78–85, which summarizes, "The outward signs on which most definitions of race are based—such as skin color and hair texture—are dictated by a handful of genes. But the other genes of two people of the same "race" can be very different. Conversely, two people of different "races" can share more genetic similarity than two individuals of the same race." The article describes the use of groups of *Alu* polymorphisms (short pieces of DNA) to clarify human genetic diversity. For further discussion of this controversy see also *Race* by Vincent Sarich and Frank Miele (2004). The complexities and uncertainties inherent in DNA profiling of ancient DNA—molecular archaeology or bioarchaeology—are described in *Blood Evidence* by Henry Lee and Frank Tirnady (2003) pp. 74–84, *The Molecule Hunt: Archaeology and the Search for Ancient DNA* by Martin Jones (2001).

[150]   *The Tarim Mummies* by J.P. Mallory and Victor H. Mair (2000) pp. 246–247, which lists it sources for the DNA information as Francalacci, P., 1995 "DNA analysis of ancient desiccated corpses from Xinjiang." *JIES* 23: 385–398; Francalacci, P., 1998 "DNA analysis of ancient desiccated corpses from Xinjiang, China: Further results." *The Bronze Age and Early Iron Age Peoples of Eastern Central Asia,* ed. V. Mair, 537–547, Washington & Philadephia: *JIES* Monograph 26. *JIES—Journal of Indo European Studies.* The struggle to get DNA testing done on the Tarim/Taklimakan mummies is also described in *The Mummy Congress* by Heather Pringle (2001) pp. 148–161. See also *The Mummies of Urumchi* by Elizabeth Wayland Barber (1999), pp. 193–194.

[151]   Quoted in *Alien Identities* by Richard Thompson ((1993), p. 303, from "Subterranean Worlds" by Walter Kafton-Minkel (1989), p. 191.

[152]   "A thousand li" is about 500 kilometers.

[153]   Quoted from the County Annals of Song-Zi Xian (Vol. 19, Miscellaneous), in: "Feidie Bai Wen Bai Da" ("Questions and Answers on UFOs") by Paul Dong (Moon Wai) (1983), translated by Gordon Creighton and published in *Flying Saucer Review (FSR),* as "Extracts from Paul Dong's . . . Questions and Answers on UFOs," Vol. 29, No.

6, 1984, pp. 16–17. Creighton comments "This case greatly resembles the many that I have translated over recent years, particularly from Brazil. And there is even a notable similarity in the way in which the victim is set down, for many of the Brazilian victims are also reported to have recovered consciousness and found themselves perched on top of quite inaccessible high places and mountain peaks." A rather nasty variant of these encounters with strange lights has emerged out of Brazil. The so-called "chupas" with their lethal beams of light have terrorised, injured and perhaps even killed locals. See *UFO Danger Zone* by Bob Pratt (1996) and *Confrontations* by Jacques Vallee (1990).

[154] Such "trips" through altered time are standard fare in "fairy" lore. See for example "Time in Fairyland" in *A Dictionary of Fairies* by Katharine Briggs" (1977) pp. 398–400, *Passport to Magonia* by Jacques Vallee (1969), *Report on Communion* by Ed Conroy (1989), and *Alien Identities* by Richard Thompson (1993). See "Paranormal and Occult Theories about UFOs" in *The UFO Encyclopedia,* 2nd Edition, by Jerome Clark, pp. 702–708, and "The Thickets of Magonia" by Jerome Clark, *International UFO Reporter,* 15, 1 (January/February 1990) pp. 4–11.

[155] "Flying saucers don't favour foreign countries only with their visits" by Zhang Ke-Tao, quoted in "Feidie Bai Wen Bai Da" ("Questions and Answers on UFOs") by Paul Dong (Moon Wai) (1983), translated by Gordon Creighton and published in *Flying Saucer Review (FSR),* as "Extracts from Paul Dong's . . . Questions and Answers on UFOs," Vol. 29, No. 6, 1984, p. 17. This alleged 1975 event from China bears a striking similarity to the controversial experience of Chilean Corporal Armando Valdes in 1977 who was alleged to have disappeared in the wake of an encounter with strange lights. Valdes also returned, but accounts claimed he had a calendar watch, which allegedly advanced five days, and had a beard growth of several days in appearance, even though he was only gone for some fifteen minutes. See for example *Time Storms* by Jenny Randles (2001), and "Comparison of two spectacular reports from Chile and China" by Gordon Creighton reprinted in *FSR* Vol. 44/2, Summer 1999, pp. 3–7. This issue contains some additional detail on Corp. Valdes' encounter. A more skeptical take on the Valdes claim appears in "The Downfall of Corporal Valdes" by Diego Zuniga, *Magonia Supplement* No. 50, May, 2004.

[156] From *Red Dust* by Ma Jian (2002), pp. 278–279.

[157] Personal interview with Ma Jian, via an interpreter, May 2003.

[158] "I Had Sex with an Alien!" by Matthew Forney, *Time,* September 29, 2003.

[159] Meng Zhao Guo is sometimes rendered as Mong and or Zhaoguo.

[160] Ibid., *Time.*

[161] A 40-page CURO report on the affair was published in their publication *Exploration of Mysteries from the Cosmos and the Earth 12/95* (1995).

[162] Sun Shi Li in an interview with Antonio Huneeus, "Ufology in the Far East—China, Japan, Korea" by J. Antonio Huneeus, *MUFON Symposium Proceeding,* 1997, p. 274.

[163] "I Had Sex with an Alien!" by Matthew Forney, *Time,* September 29, 2003.

[164] Antonio Huneeus, "Ufology in the Far East—China, Japan, Korea" by J. Antonio Huneeus, *MUFON Symposium Proceeding,* 1997, p. 274. There may be some confusion here as Sun Shu Li indicates in his Australian lecture via a translator that sexual relations occurred in Zhao Guo's farm house.

[165] My transcript from the lecture is quoted with the permission of the conference organiser, Glennys Mackay.

[166] Sun Shi Li's lecture was in Spanish. The translator's version is given. Sun Shi Li acted as a foreign language translator for Spanish. He does not speak good English. Because of the translation process parts of the material are fragmented, and may be subject to correction.

[167] From "Focus: Ufology in Mysticism," *China Daily Hong edition,* December 12, 2003.

[168] An entertaining account of the Basque people can be found in *The Basque History of the World* by Mark Kurlansky (2000).

[169] C. L. Turnage refers to the unverified claims of Dan Sherman, recounted in his book *Above Black* (1997), where he describes "insider" claims about "alien interbreeding with man" indicating "the Basque people of Spain and France are the most likely candidates in the search for their progeny. The Basque language has no identifiable roots, and they are genetically different from all other humans on the planet. The scientific community considers them a human anomaly." See "Sexual Encounters with Extraterrestrials" by C. L. Turnage

(2001) p. 117. This is a rather resonating but unverified connection to my own comments on "the Basque Legacy."

[170] From "Tenacity in Religion, Myth, and Folklore: The Neolithic Goddess of Old Europe Preserved in a Non-Indo-European Setting" by Michael Everson, *Journal of Indo-European Studies* 17 Nos. 3 & 4, Fall/Winter 1989, pp. 277–95.

[171] Perhaps an interesting place to contemplate that Mt. Perdido in the Spanish Pyrenees, near the Basque territories, was one of four *underground* "UFO bases" identified by remote viewing back in the 1970s by pioneer and "star" remote viewer Pat Price. For further details see the footnoted comments re Mt. Inyangani in the chapter on Credo Mutwa.

[172] The claim of a "copy" or "clone" being produced by aliens appears in a number of cases. For example the 1975 Brazilian case of Antonio Alves Ferreira featured the production of a somewhat "flawed" copy. It allegedly lacked a scar on the base of the foot and seemed to be very much heavier than Ferreira. For the bizarre and rather unbelievable details see "The Case of Antonio Alves Ferreira" by Irene Granchi, *FSR*, Vol. 31, No. 2, 1986, pp. 6–12. Irene Granchi indicates, "I know of other cases in which a person has seen such a duplication performed, and has seen himself, walking around in his own home, and not even spotted as a duplicate by his own wife! While in the meantime, the original human being in question was still aboard the flying saucer!" (p. 10)

[173] "The Jinn and the dolmen: the Most Amazing Case of Abduction Yet" by Antonio Riberia, *FSR*, Vol 31, No. 4, 1986, pp. 2–12.

[174] *Time Storms* by Jenny Randles (2001) provides an interesting interpretation of such cases.

[175] See "Australian 'Interrupted Journeys'" by Bill Chalker, *MUFON UFO Journal*, No. 150, August 1980, and *The OZ Files* by Bill Chalker (1996).

[176] "Road Hazard Down Under?" by Bill Chalker, *FSR*, Vol. 22, No. 5, 1976, pp. 31–32.

[177] "Road Hazard Down Under?" by Bill Chalker, *FSR*, Vol. 22, No. 5, 1976.

[178] Communication from Vicente-Juan Ballester Olmos, January, 2004.

[179] "Alleged Experiences Inside UFOs: An Analysis of Abduction Reports," *Journal of Scientific Exploration,* Vol 8, No 1, pp. 91–105, 1994.

[180] *Oxford Dictionary of Celtic Mythology* by James Mackillop (1998) pp. 414–416.

[181] Ibid. Mackillop, p. 386.

[182] The UFO literature and the debate includes the following material: *Passport to Magonia* by Jacques Vallee (1969), *Report on Communion* by Ed Conroy (1989) particularly chapters 7 and 8: "Little Green Men: From Ireland—or the Pleiades?" and "The Green Men, the Goddess, and the Fairy Queen," and *Alien Identities* by Richard Thompson (1993), particularly chapter 8: "Modern Observations and Ancient Traditions." Researchers such as Jerome Clark regard these sorties into "fairy lore" and uncertain folklore as untenable and damaging to the credibility of serious UFO research. See his articles "Paranormal and Occult Theories about UFOs" in *The UFO Encyclopedia,* 2nd Edition, pp. 702–708, "The Thickets of Magonia," *International UFO Reporter,* 15, 1 (January/February 1990) pp. 4–11, and "Fairies" pp. 114–122 in *Unexplained* (1993). Further research and debate is needed in this controversial area. A somewhat more erratic fingering of the Tuatha and "the Underworld," as our alien "suspects" can be found in *Architects of the Underworld* by Bruce Rux (1996). By the 1970 edition of his collaboration with George Adamski, *Flying Saucers Have Landed,* the delightfully eccentric Desmond Leslie suggests, "Celtic scholars might do well to re-examine the Tuatha de Danaan legends; that semi-mythical magical race of 'gods' who descended from the sky and made their headquarters on and around Lough Erne, County Fermanagh; people of fair skin, red-golden hair, and green eyes. Their collection of magical gadgetry has all the ring of a highly scientific space expedition, observed by awestruck natives who worshipped them as gods." (p. 166) Leslie was a bit creative with his Tuatha descriptions but red to fair to blond hair, and from blue to green eyed, has been embraced in the Tuatha literature.

The Tuatha and fairy lore literature is massive: It is fair to say that there are more differences between fairy lore accounts and alien abduction accounts than there are similarities. Therein lies the quandary—a connection or a damaging distraction? *The Fairy-faith in*

Celtic Countries by W.Y. Evans-Wentz (1911, 1977) is a classic on the topic. See also for example *Robert Kirk: Walker Between Worlds* (a new edition of *The Secret Commonwealth of Elves, Fauns and Fairies* compiled by R. J. Stewart (1990), *Oxford Dictionary of Celtic Mythology* by James Mackillop (1998), *Irish Myths and Legends* by Lady Gregory, *A Dictionary of Fairies* by Katharine Briggs (1977), *The Candle of Vision* by Æ (George Russell), *Fairies* by Janet Bord (1997), and *Troublesome Things* by Diane Purkiss (2000).

[183] *The Fairy-faith in Celtic Countries* by W.Y. Evans-Wentz (1911, 1977) pp. 83–84.

[184] *The Candle of Vision* by Æ (George Russell) (1965), pp. 94–97.

[185] Interpretations of what all this means are diverse. For example see the work of Paul Devereux, in particular, *Earth Lights* (1982) and *Earth Lights Revelation* (1989). See also *UFOs and Ufology* by Paul Devereux and Peter Brookesmith (1997), particularly Chapter 10: "Planet Earth's UFOs—Chasing the Wild Lights." Other researchers provide different views, for example *Alien Energy* by Andrew Collins (2003) and *Electric UFOs* by Albert Budden (1998). Excellent studies of particular areas can be found in *Supernatural Peak District* by David Clarke (2000) and *Supernatural Pennines* by Jenny Randles (2002). David Clarke also wrote with Andy Roberts *Twightlight of the Celtic Gods* (1996). See also *The Pennine UFO Mystery* by Jenny Randles (1983).

[186] Personal communication. The deity details were confirmed in James Mackillop's *Oxford Dictionary of Celtic Mythology* (2000), p. 387.

[187] This bizarre nexus is potently revealed in the shamanic vision experienced by anthropologist Dr. Michael Harner. He spent over a year with the Conibo Indians of the Ucayali River region in the Peruvian Amazon during 1960 and 1961. He found that local shamans accessed the "shamanic realm" by using a sacred drink prepared from "ayahuasca" (the "soul vine," "the vision vine," "the little death"). Harner took the drink with dramatic results. He experienced a remarkable shamanic vision. He found himself in something resembling a huge fun house, "a supernatural carnival of demons," presided over by "a gigantic, grinning crocodilian head, from whose cavernous jaws gushed a torrential flood of water." The flood transformed the scene into "a simple duality of blue sky above and sea below." He saw two strange boats wafting back and forth in the air

and coming closer. They slowly combined to form a single vessel like a Viking ship.

This detail of two "ships" or aerial craft coming together and blending into one was bizarrely played out in the extraordinary UFO encounter of Dr. X in France during 1968. The craft projected beams of light that may have been linked to remarkable "rapid healing" of an injury the witness had.

"Ethereal singing" could be heard and he was able to make out "bird-headed" humans onboard. At the same time some sort of "essence" began to float up from his chest into the boat. A form of paralysis started to set in. Fearing he was dying, Harner tried to summon help from the Indians that were around him in reality but not accessible in this shamanic state of consciousness. Harner gradually became aware of more visions and information, being presented to him by "giant reptilian creatures reposing sluggishly at the lowermost depths of the back of my brain. I could only vaguely see them in what seemed to be gloomy, dark depths." He was given information reserved for the dying and the dead, he was told.

The "reptiles" projected a visual scene in front of Harner, which showed the creation of life on earth. But first before that life there were hundreds of "large, shiny, black creatures with stubby pterodactyl-like wings and huge whale-like bodies" that dropped from the sky and landed on the barren landscape. Their heads were not visible. These beings were fleeing from enemies from outer space. They created life on earth "in order to hide within the multitudinous forms and thus disguise their presence."

Harner learned that "the dragon-like creatures were thus inside of all forms of life including man [almost like DNA, although at the time, 1961, Harner knew nothing of DNA]. They were the true masters of humanity and the entire planet," the "dragon-like creatures" told Harner. He was caught up in a struggle being the aerial ship with the "bird" people taking his "soul" and the "dragon-like denizens of the depths." As he felt he was about to die Harner was able to cry out "medicine," whereupon in the real world the Indians frantically set about making an antidote. A powerful being appeared before Harner to protect him from "the alien reptilian creatures."

Simultaneously the antidote to "the little death" drink was given, and the dragons and "the soul boat" disappeared. While the antidote eased his condition, Harner continued having "many additional visions of a more superficial nature . . . I made fabulous journeys at will through distant regions, even out into the galaxy; created incredible architecture; and employed sardonically grinning demons to realize my fantasies." He finally slept until waking surprisingly refreshed and peaceful the following morning.

Harner found that he could remember all of the experience except for "the communication from the dragon-like creatures." Eventually he was able to remember it. He was then seized by a fear that he should not know this material, that it seemed intended only for the dying. His remedy—he quickly told others, including local missionaries, who were startled by the similarities with parts of the Book of Revelations. This came as a surprise to Harner—an atheist. Harner also sought advice on his "vision" from "the most supernaturally knowledgeable of the Indians, a blind shaman who had made many excursions into the spirit world with the aid of the "ayahuasca" drink. The shaman said of Harner's "masters of the earth," "Oh, they're always saying that. But they are only the Masters of Outer Darkness." The shaman further surprised Harner when he casually indicated that these "spirits" came from the sky. Harner had not yet mentioned that he had seen them coming from outer space! There are numerous correspondences with modern-day abduction narratives.

See Harner's book *The Way of the Shaman* (1980), pp. 1–8. Harner's account is also described in *The Cosmic Serpent* (1999) an intriguing book by Jeremy Narby. See also "I felt like Socrates accepting the Hemlock" by Michael Harner in *Shamans through Time: 500 Years on the Path to Knowledge*, edited by Jeremy Narby and Francis Huxley (2001). Narby also elaborates the possible latent chemical "knowledge" accessed via shamanic "visions." Before automatically dismissing such things, as a chemist myself, I often reflect on chemists Kekule and Mendeleyev's visionary dreams that led to the elaboration of the structure of benzene and the periodic table of elements respectively. The elements of ayahuasca mediated DNA visions and shamanic "alien" visions and encounters can be further considered in *Ayahuasca Visions* (1991) by Luis Eduardo Luna and Pablo Amaringo, and in "Experiences of Light and Balls of Fire" in Holger Kalweit's *Dreamtime and Inner Space* (1988).

[188] Blind faith in the infallibility of DNA profiling is an area of growing concern. Usually the basis of these concerns is either contamination, intentional or unintentional, or problems in methodology. Australian molecular biologist Dr. David Berryman, the Managing Director of the Western Australian State Agricultural Biotechnology Centre at Murdoch University, has been drawing attention to the potential of deliberate contamination of crime scenes with PCR products:

In an interview on *The Big Idea*, Radio National ABC, September 22, 2002, Dr. Berryman indicated, "The most common way of doing the forensic analysis at the moment is using PCR amplification of variable regions of DNA, called 'short hand and repeats' or STRs, but with forensics in particular, you often work with very small amounts of material—it might be a single hair or even the DNA contained in the cells on a fingerprint on a stamp, or something like that—so you've really got very little DNA, there's not really enough there to do any analysis on, somehow you've got to increase the amount, and that's done by PCR. But once again, PCR has contamination issues that you have to be aware of in the laboratory, but those issues are still out there in the field, too. And if people want to contaminate a crime scene, using PCR as the diagnostic test is a bit problematic because, as I said, you can contaminate a crime scene using PCR products. You don't have to distribute hairs or try to put a fingerprint on a wall in the space that looks like it was accidentally put there during a crime—you can just cover the whole scene with DNA PCR products."

Dr. Berryman explained the process: "The two steps would be; firstly, get access to the kit and amplifier, and that's easy to do, it's just publicly available information—it's like going down to the supermarket and buying a litre of milk, it's not hard to do; then, do the PCR reaction and once again, that's easy to do, it's all publicly available information, it doesn't require any skill or any significant resource—in fact, you can do it at home on your stove. Then just go to a crime scene and squirt it around, and that's enough to contaminate a crime scene from our data, virtually beyond any possibility of recovering a genuine profile." What concerned Berryman was that, while there was plenty of effort in resolving such PCR-related contamination issues in a lab environment, the impact of deliberate PCR contamination in the field, i.e., at crime scenes, seemed largely ignored.

Source: http://www.abc.net.au/rn/bigidea/stories/s673661.htm. See also "Shadows of Doubt" by Victoria Laurie, *The Weekend Australian Magazine*, September 27–28, 2003.

Before someone suggests it, the chances of Peter Khoury having this sort of knowledge, only now being discussed and debated in the field of forensics, were certainly nil back in 1992 and pretty much the same even now. The dilemma now would be how anyone could fabricate a PCR "brew" in a similar way that would contain the same rare genetic substitutions. Impossible? No, but very difficult, given the rarity of the DNA profile and its unusual attributes near the root and at the shaft of the hair. Sadly, given the propensity for hoaxing in this field it is only a matter of time before someone tries it. Thus analysts and researchers working in this area need to rigorously verify context and chains of evidence in such events.

[189] *Life's Solution* by Simon Conway Morris (2003). See also *Rare Earth* by Peter Ward and Donald Brownlee (2000), and "Rare Earths and Hidden Agendas" in *Life Everywhere* by David Darling (2001). See also *The Extraterrestrial Encyclopedia* also by David Darling (2000).

[190] The "Life Principle" is evocatively argued in *Planetary Dreams* by Robert Shapiro (1999).

[191] "Do We Have to Spell It Out?" by Paul Davies, *New Scientist*, August 7, 2004.

[192] *Pale Blue Dot* by Carl Sagan (1995).

[193] *That Eye, the Sky* by Tim Winton (1986).

[194] Other worthwhile contributions to the extraterrestrial life debate can be found in such books as *The Biological Universe* by Steven Dick (1996), *The Fifth Miracle* by Paul Davies (1998) (reissued & revised as *The Origin of Life* (2003)), *There Be Dragons* by David Koerner and Simon LeVay (2000), *Our Cosmic Origins* by Armand Delsemme (1998), *Many Worlds* edited by Steven Dick (2000), *Sharing the Universe* by Seth Shostak (1998), *Evolving the Alien* by Jack Cohen and Ian Stewart (2002), *Where Is Everybody?* by Stephen Webb (2002), and *Lonely Planets* by David Grinspoon (2003).

[195] Hopkins, Budd (1987) *Intruders: The Incredible Visitations at Copley Woods* (Random House, New York).

[196] Mack, John E. (1994) *Abductions: Human Encounters with Aliens* (Simon & Schuster, London).

[197] Chalker, B. (1996) *The Oz Files: the Australian UFO Story* (Duffy & Snellgrove, Potts Point, Australia).

[198] Wilson, M. R., et.al. (1995) "Extraction, PCR amplification and sequencing of mitochondrial DNA from human hair shafts." *Biotechniques* 18, 662–669.

[199] Krings, M., et.al. (1997). "Neanderthal DNA sequences and the origin of modern humans." *Cell* 90, 19–30.

[200] This technical report by the principal biochemical scientist involved with the APEG was part of a 15 page article I prepared entitled "Strange Evidence" which appeared in the Spring 1999 issue of the J. Allen Hynek Center for UFO Studies publication *IUR—the International UFO Reporter*. I was pleased that it appeared in that publication because it had to run the gauntlet of some good solid peer review, particularly from Dr. Michael Swords, Department of Science Studies, Western Michigan University, and Dr. Mark Rodeghier, University of Illinois at Chicago. The initial publication drew considerable interest, even from a number of biochemists. The only really critical response was drawn from someone whose e-mail name and credentials we were not able to confirm. He launched into a long convoluted run of e-mails claiming there were many errors in the article. We asked him to elaborate. Although he never supplied the full list of "errors," those he did elaborate on turned out to be trivial, misguided, or inappropriate. We initially responded in good faith despite the litany of largely spurious "errors," but soon the line of questioning became very misguided. He construed the mention of a Chinese man in our research report, who we DNA tested to rule out the remote possibility of contamination, in lieu of the rare Asian Mongoloid DNA present in the hair sample, was evidence that his "Chinese man of mystery" had been to the Khoury home and had sex with both Peter and Vivian Khoury. This very bizarre leap in logic on the part of this critic led us to finally give up trying to respond to the increasingly irrelevant, misguided, and incorrect error claims. The critic's "Chinese man of mystery" was no real mystery. He was a lab worker—a postgraduate in biochemistry—who

had been part of our analytical team. Even though we believed that he had not come into direct contact with the hair sample, given that he was in the lab where we had stored the sample for a while, we felt we needed to rule his DNA in or out as the possible source of the unusual DNA found. It turned out his DNA was as expected, normal Chinese DNA with no rare or unusual aspects. He had never met Peter or Vivian Khoury and had never been to their home. Such convoluted, bizarre, and spurious arguments were a prominent part of the critic's attack on the work we had done. More rational reviews have confirmed that the research is reasonable and worth serious consideration. For those who wish to follow this torturous debate you can do so by consulting the skeptically oriented UFORL—UFO Research List on the Internet at http://groups.yahoo.com/group/UFORL/message/ with the messages covering the period October to November 2001. It may help you to consider my final response first: Message 338 dated November 17, 2001. My responses to the issue are included as messages 262, 269, 283, 305, and 338. The critic's messages are messages 256, 268, 276, 303, 319, 329, 332, 346. Other list member responses include messages 286, 280, and 278.

[201] Horai, S., and Hayasaka, K. (1990). "Intraspecific nucleotide sequence differences in the major noncoding region of human mitochondrial DNA." *American Journal of Human Genetics* 46, 828–842.

[202] Horai, S., et.al. (1996). "MtDNA polymorphism in East Asian populations." *American Journal of Human Genetics* 59, 579–590.

[203] Jorde, L. B., et.al. (1998). "Using mitochondrial and nuclear DNA markers to reconstruct human evolution." *Bioessays* 20, 126–136.

[204] The putative "yowie" hair was acquired from Peter Khoury who received it from a young Australian person who had reported encounters with a strange creature. That witness put the encounter in a UFO context. The analyst used the identifier of "yowie," the legendary Australian bush creature akin to the American Bigfoot. A closer examination of the young witness' description makes the Yowie identification inappropriate. The sample yielded only weak human consensus amplification suggesting possible contamination, thus while not satisfactory in its own right it served the purpose of a secondary "control sample."

[205] Two different samples from the hair root gave two different DNA sequences, within the same PCR: 162555 genotype for 6/7 clones from the soft root (i.e., the skin area of the "donor") and 16, 304 again (as in the shaft result from the first round of testing) from the hard keratin inside the base of the hard shaft of the hair). This was not evidence of heteroplasmy (where two different mitochondrial DNAs rarely appear within the sample, usually a result of co-existence of mutant mitochondrial and "wild-type" DNA molecules within a cell or tissue—See *Human Molecular Genetics,* Strachan and Read, 1997, p. 592). Heteroplasmy, which is more readily found in human hair than other parts of the body, refers to single-base transitions in the mitochomdrial DNA. For example G to A, or one C to T. They are not big changes. Environmental exposure and aging can be factors. In the Khoury samples the different sequences were striking and rare. The principal biochemist suggests that this may be evidence for advanced DNA cloning techniques, with the blond female possibly having hair replacement in the past. The *Nature Biotechnology* paper highlights very recent advances that are occurring in baldness research that may be analogues of the apparent procedures evident in the strange hair sample. The other remote alternative is that the hair sample was handled forcefully by someone with an extremely rare C-to-T DNA substitution at 16,108. Given the circumstance of the recovery of the hair sample there appear to be only two candidates—the strange blond female or the unusual "Asian" female. This possibility still supports the claim that Peter Khoury was in the presence of two very unusual persons with rare mitochondrial DNAs, not easily found in the Sydney, Australia, region. The results do not support prosaic processing errors or contamination.

ACKNOWLEDGMENTS

Thank you Anne, Kieran, and Christopher for your ongoing interest and support in this project.

I would like to thank Peter Khoury for his courage, determination, and cooperation in the demanding research and investigation that led to this book. I thank him and his family (Vivian, Stephen, and Georgia) for their support. I also thank the following people, who provided information or assistance in this investigation: Sam Khoury, Frankh Wilkes, Jamie Leonarder, Aspa Vetsikes, Moira McGhee, Bryan Dickeson, Frank Sinclair, and in particular, Robb Tilley and Dick Warburton for their help, support, and friendship throughout this investigation and many others. There are others who do not wish to be identified. In particular I would like to thank the principal biochemist and the other scientists who worked on the PCR DNA investigation of the Khoury hair sample and produced the intriguing data that is so much a part of this story, and certainly a potent catalyst for the forensic and scientific paradigm that needs to be a vital part of the research into the alien abduction controversy. In addition I would like to express my deepest appreciation of the wonderful support given to me by one particular party who does not wish to be identified. Without that ongoing support, this continuing research program would not have been possible.

Thanks also to Diane Harrison for her investigative and research support and her fine contributions to this field, particularly with the Australian UFO Research Network (AUFORN). Her partner, Robert Frola, the editor of the *Australasian Ufologist*, also was very supportive of this research.

Thanks to Kary Mullis for his information and interview, and of course for his remarkable PCR concept that ultimately transformed the field of biochemistry. Thanks for the invitation to relax with a beer and "watch the boats go by" some day in California, but just the same I think I'd like to hang out for a while at his Mendocino county property to "racoon" hunt, if the opportunity arises.

My appreciation and thanks go to the rest of those who have contributed their time and their experiences: Kelly Cahill, "Mike Wood" (and to Mark Nolan for his assistance in that inquiry), Gary Lowrey, Vusamazulu Credo Mutwa, Ma Jian, and the others who did not wish to be identified.

My thanks to the following people who assisted my diverse investigations in various ways: Budd Hopkins, John Mack, Dominique Callimanopulos, Whitley Strieber, Phyllis Budinger, Colm Kelleher, and Tom Deuley for U.S. and general inquiries; Vicente-Juan Ballester Olmos for Spanish inquiries; Jenny Randles, David Clarke, and Dave Baker (also for being great guides), and Andy Roberts for inquiries in England; Harry Turner, Marion Leiba, Gillian Summers, Wayne Medway, Ron Summers, Gerry Klein, and Adam Klein for their information about a special lady, Vicki Klein; Noi for the Thai Naga reveries; Sophia Ya Fei, "Jian," and others for Chinese hospitality and perspectives; Lee for an alternative focus; Rick Martin for permission to use parts of his interview with Credo Mutwa; Brian (Baruch) Crowley for his recollections of Credo Mutwa; Robert Lawlor for alternative perspectives; Robert Maragna for inquiries into the Gundiah affair; Keith Basterfield for his review of implants; Roger Leir for his implant work and inquiries into the Gary Lowrey story; Steve Walters, Rodney Coutts and Dudley Robb for discussions and assistance on some sources; and to the many others who have helped along the way.

By way of estranged thanks, I also would love to hear from Keith, Amy, and Petra to understand what really happened on that dark and stormy night in Gundiah and on the journeys to and from Mackay.

Many thanks to my editor, Patrick Huyghe, who encouraged me to tell this story with its forensic perspective on "the UFO beat."

Bill Chalker
P.O. Box 42, West Pennant Hills, NSW, 2125, Australia
www.theozfiles.com
www.bananatv.com/ufo

Forensics and science

Bamshad, Michael and Olson, Steve. "Does race exist?" *Scientific American,* December 2003.

Barber, Elizabeth Wayland. *The Mummies of Urumchi,* 1999.

Barsh, G. "What Controls Variation in Human Skin Color?" *PloS Biology,* 1/1:19–22, 2003.

Bova, Ben. *The Story of Light,* 2001.

Bremner, Moyra. *GE: Genetic Engineering and You,* 1999.

Cavelos, Jeanne. *The Science of the X-Files,* 1998.

Clegg, Brian. *Light Years: An Exploration of Mankind's Enduring Fascination with Light,* 2001.

Cohen, Jack, and Ian Stewart. *Evolving the Alien: The Science of Extraterrestrial Life,* 2002.

Darling, David. *The Extraterrestrial Encyclopedia: An Alphabetical Reference to All Life in the Universe,* 2000.

Darling, David. *Life Everywhere: The Maverick Science of Astrobiology,* 2001.

Davies, Paul. *Are We Alone?* 1995.

Davies, Paul. *The Fifth Miracle: The Search for the Origin of Life,* 1998 (reissued and revised as *The Origin of Life* 2003).

Delsemme, Armand. *Our Cosmic Origins: From the Big Bang to the Emergence of Life and Intelligence,* 1998.

Dick, Steven. *Life on Other Worlds: The 20th Century Extraterrestrial Life Debate,* 1998.

——. *The Biological Universe,* 1996.

——, ed. *Many Worlds: The New Universe, Extraterrestrial Life and the Theological Implications,* 2000.

Dennis, Carina, and Richard Gallagher, eds. *The Human Genome,* 2001.

Erzinclioglu, Dr. Zakaria. *Every Contact Leaves a Trace: Scientific Detection in the Twentieth Century,* 2000.

Evans-Wentz, W.Y. *The Fairy-Faith in Celtic Countries,* 1911, 1977.

Fox, Douglas. "Keep Your Hair On: How Gene Therapy and Cloning Could Make It Grow Again." *New Scientist,* October 13, 2001, 28–35.

Grinspoon, David. *Lonely Planets: The Natural Philosophy of Alien Life,* 2003.

Hauu, et al. "Observation of Coherent Optical Information Storage in an Atomic Medium Using Halted Light Pulses." *Nature,* January 25, 2001.

Hoffman, R. "The Hair Follicle as a Gene Therapy Target." *Nature Biotechnology,* Vol. 18, January 2000, 20–21.

Jones, Martin. *The Molecule Hunt: Archaeology and the Search for Ancient DNA,* 2001.

Kaku, Michio. *Hyperspace: A Scientific Odyssey through Parallel Universes, Time Warps, and the 10th dimension,* 1994.

——. "Playing God: Designer Children and Clones." In *Visions: How Science will Revolutionize the 21st Century and Beyond,* 1998.

Koerner, David, and Simon LeVay. *There Be Dragons: The Scientific Quest for Extraterrestrial Life,* 2000.

Laurie, Victoria. "Shadows of Doubt." *The Weekend Australian Magazine,* September 27–28, 2003.

Lane, Brian. *The Encyclopedia of Forensic Science,* 1992.

Lee, Dr. Henry, and Frank Tirnady. *Blood Evidence: How DNA Is Revolutionizing the Way We Solves Crimes,* 2003.

Leroi, Armand Marie. *Mutants: On the Form, Varieties and Errors of the Human Body,* 2003.

Mallory, J.P., and Victor H. Mair. *The Tarim Mummies: Ancient China and the Mystery of the Earliest Peoples from the West,* 2000.

Morris, Simon Conway. *Life's Solution: Inevitable Humans in a Lonely Universe,* 2003.

Mullis, Kary. "The Unusual Origin of the Polymerase Chain Reaction" *Scientific American,* April 1990.

——. *Dancing Naked in the Mind Field,* 1998.

Oberg, Leon. "Earthquake Alert," *Sydney Sun-Herald,* August 20, 1989.

Owen, David. *Hidden Evidence,* 2000.

Platt, Richard. *Crime Scene: The Ultimate Guide to Forensic Science,* 2003.

Pringle, Heather. *The Mummy Congress: Science, Obsession, and the Everlasting Dead,* 2001.

Rabinow, Paul. *Making PCR: A Story of Biotechnology,* 1997.

Ramsland, Katherine. *The Forensic Science of C.S.I.,* 2001.

Rees, J. L. "The Melancortin 1 Receptor (MC1R): More Than Just Red Hair." *Pigment Cell Research,* 13: 135–140, 2000.

Reynolds, A.J., et al. "Trans-gender Induction of Hair Follicles." *Nature,* Vol. 402, November 1999.

Sagan, Carl. *Pale Blue Dot: A Vision of the Human Future in Space,* 1995.

——. *The Demon-Haunted World: Science as a Candle in the Dark,* 1996.

Shapiro, Robert. *Planetary Dreams: The Quest to Discover Life Beyond Earth,* 1999.

Shostak, Seth. *Sharing the Universe: The Quest for Extraterrestrial Life,* 1998.

Silver, Lee M. *Remaking Eden: Cloning and Beyond in a Brave New World,* 1998.

Simon, Dr. Anne. *The Real Science behind the X-Files,* 1999.

Strachan, Tom, and Andrew Read. *Human Molecular Genetics,* Oxford, BIOS Scientific, 1997.

Sturm, R.A., N.F. Box, and M. Ramsay. "Human Pigmentation Genetics: The Difference Is Only Skin Deep." *Bioessays,* 20: 712–721, 1998.

Talbot, Michael. *The Holographic Universe,* 1991.

Tributsch, Helmut. *When the Snakes Awake: Animals and Earthquake Prediction,* 1982.

Ward, Peter, and Donald Brownlee. *Rare Earth: Why Complex Life Is Uncommon in the Universe,* 2000.

Watson, James, with Andrew Berry. *DNA: The Secret of Life,* 2003.

Webb, Stephen. *Where Is Everybody? Fifty Solutions to the Fermi Paradox and the Problem of Extraterrestrial Life,* 2002.

White, Michael. *The Science of the X-Files,* 1996.

Wilson, Clare. "Ready for Your Close-up?" *New Scientist,* July 20, 2002.

Wilson, M.R., D. Polanskey, J. Butler, J. DiZinno, J. Replogle, and B. Budowle. "Extraction, PCR Amplification and Sequencing of Mitochondrial DNA from Human Hair Shafts." *BioTechniques,* Vol. 18, No. 4, 662–669.

UFO material

Alexander, Victoria. "New Protocol for Abduction Research," *MUFON UFO Journal,* No. 307, November 1993.

Anon. "Focus: Ufology in Mysticism." *China Daily Hong edition,* December 12, 2003.

Ashpole, Edward. *The UFO Phenomena: A Scientific Look at the Evidence for Extraterrestrial Contacts,* 1996.

Auchettl, John. "The Tyers Farm Event: Invisible Boundary Window." *UFO Quest,* Issue #1, October 2003.

Ballester Olmos, Vicente-Juan, and Jacques Vallee. "Type-1 Phenomena in Spain and Portugal: A Study of 100 Iberian Landings." In "UFOs in Two Worlds," *FSR Special Issue,* No. 4, August 1971.

——. *Ovnis: El Fenomeno Aterrizaje (UFOs: The Landing Phenomenon),* 1978.

——. "Alleged Experiences Inside UFOs: An Analysis of Abduction Reports." *Journal of Scientific Exploration,* Vol 8, No 1, 1994.

Basterfield, Keith. *UFOs: A Report on Australian Encounters,* 1997.

——. "The Implant Motif in UFO Abduction Literature." *Journal of UFO Studies,* New Series Vol. 8, 2003.

Bowen, Charles, ed. *The Humanoids,* 1969.

Brookesmith, Peter. *Alien Abductions,* 1998.

Bryan, C.D.B. *Close Encounters of the Fourth Kind: a Reporter's Notebook on Alien Abduction, UFOs, and the Conference at M.I.T.,* 1995.

Budden, Albert. *Electric UFOs—Fireballs, Electromagnetics and Abnormal States,* 1998.

Budinger, Phyllis. "Betty Hill Dress Analysis." *Technical Service Response* No. UT025, October 19, 2003.

Bullard, Thomas E., PhD. *On Stolen Time: A Summary of a Comparative Study of the UFO Abduction Mystery,* 1987.

Bullard, Thomas E. "Abduction Phenomenon." In Jerome Clark, *The UFO Encyclopedia: The Phenomenon from the Beginning,* 1998.

Carpenter, John. "Blinded by the Light: An Analysis of Missing Time Captured on Videotape." *MUFON Journal* No. 373, May 1999 and *2000 MUFON Symposium proceedings.*

——. "Analysis Supports Abduction on Video." *MUFON Journal* No. 390, October 2000.

Chalker, Bill. "Road Hazard Down Under?" *FSR,* Vol. 22, No. 5, 1976.

——. "Australian 'Interrupted Journeys,' " *MUFON UFO Journal,* No. 150, August 1980.

——. "Physical Traces." In *UFOs 1947–1987: The 40-year Search for an Explanation,* edited by Hilary Evans with John Spencer, 1987.

——. "An Extraordinary Encounter in the Dandenong Foothills." *International UFO Reporter,* September/October, 1994.

——. *The OZ Files: The Australian UFO Story,* 1996.

——. "Strange Evidence." *International UFO Reporter (IUR),* Spring, 1999.

——. "UFO Abductions & Science: A Case Study of Strange Evidence." *Australasian Ufologist,* Vol. 3, No. 3, 3rd Quarter, 1999.

——. "The World's First DNA PCR Investigation of Biological Evidence from an Alien Abduction." *Hard Evidence,* Vol. 1 No. 2, January/February 2001.

Chalker, Bill, and Diane Harrison. "The Gundiah Mackay Abduction Milieu: A Preliminary Report," *Australasian Ufologist,* Vol. 5 No. 4, 2001.

Clark, Jerome. "The Thickets of Magonia." *International UFO Reporter,* 15, 1, January/February 1990.

——. *The UFO Book,* 1997.

——. *The UFO Encyclopedia,* 2nd edition, 1998.

——. "Close Encounters of the Third Kind." In *UFO Encyclopedia,* Vol. 1, 1998.

——. "Hill Abduction Case." In *UFO Encyclopedia,* Vol. 1, 1998.

——. "Paranormal and Occult Theories about UFOs." In *UFO Encyclopedia*, Vol. 2, 1998.

——. "Villas-Boas CE3." In *UFO Encyclopedia*, Vol. 2, 1998.

Collins, Andrew. *Alien Energy: UFOs, Ritual Landscapes and the Human Mind*, 2003.

Connelly, Dwight (compiler). *The World's Best UFO Cases*, 2004.

Conroy, Ed. *Report on Communion: An Independent Investigation and Commentary on Whitley Strieber's Communion*, 1989.

Corcoran, Kevin. "Why Did They Do It?" "And Where Are They Now? Beam of Light Carries Woman through Window to 1.8 m Alien," "If a Hoax: Police Say It Was: What Reason?" "Mayor Reckons Affair Certainly Was No Hoax," and "Researchers Find Story Did Not Add Up." *Fraser Coast Chronicle*, October 11–12, 2002.

Creighton, Gordon. "The Amazing Case of Antonio Villas Boas." In *The Humanoids*, ed. Charles Bowen, 1969.

——. "Comparison of Two Spectacular Reports from Chile and China." Reprinted in *FSR* Vol. 44/2, Summer 1999.

Dean, Jodi. *Aliens in America: Conspiracy Cultures from Outerspace to Cyberspace*, 1998.

Dennett, Preston. *UFO Healings: True Accounts of People Healed by Extraterrestrials*, 1996.

Denzler, Brenda. *The Lure of the Edge: Scientific Passions, Religious Beliefs, and the Pursuit of UFOs*, University of California Press, 2001.

Devereux, Paul. *Earth Lights: Towards an Understanding of the UFO Enigma*, 1982.

——. *Earth Lights Revelation: UFOs and Mystery Lightform Phenomena: The Earth's Secret Energy Force*, 1989.

Devereux, Paul, and Peter Brookesmith. *UFOs and Ufology: The First 50 Years*, 1997.

Dong, Paul (Moon Wai). "Feidie Bai Wen Bai Da" ("Questions and Answers on UFOs") 1983, translated by Gordon Creighton and published in *Flying Saucer Review (FSR)*, as "Extracts from Paul Dong's . . . Questions and Answers on UFOs." Vol. 29, No. 6, 1984.

Druffel, Ann. *How to Defend Yourself against Alien Abduction*, 1998.

——. *Firestorm: Dr. James E. McDonald's Fight for UFO Science*, 2003.

Emmons, Charles, Ph.D. *At the Threshold: UFOs, Science and the New Age*, 1997.

Fontes, Olavo, M.D., and João Martins. "Report on the Villas-Boas Incident" translated by Irene Granchi." In *Flying Saucer Occupants,* by Jim and Coral Lorenzen, 1967.

Forney, Matthew. "I Had Sex with an Alien!" *Time,* September 29, 2003.

Fowler, Raymond. *The Andreasson Affair,* 1979.

——. *The Andreasson Affair: Phase Two,* 1982.

——. *The Watchers,* 1990.

——. *The Allagash Abductions: Undeniable Evidence of Alien Intervention,* 1993.

——. *The Watchers II,* 1995.

——. *The Andreasson Legacy,* 1997.

Fuller, John G. *The Interrupted Journey: Two Lost Hours "Aboard a Flying Saucer,"* 1966.

Galindez, Oscar. "The Anthropomorphic Phenomena at Santa Isabel," by Parts 1 and 2, *FSR,* Vol. 21, No. 2, 1975, and Vol. 21 No. 3 and 4 (Double issue), 1975.

——. "The Anthropomorphic Entity at Villa Carlos Paz," *FSR,* Vol. 26, No. 5 and 6, and Vol. 27, No. 1.

Good, Timothy. *Alien Base,* 1998.

——. *Unearthly Disclosure,* 2001.

Granchi, Irene. "Abduction at Botucatu." *FSR* 30/1, 1984.

——. "The Case of Antonio Alves Ferreira." *FSR,* Vol. 31, No. 2, 1986.

Hall, Richard. *The UFO Evidence, Volume II: A Thirty Year Report,* 2001.

Hind, Cynthia. *UFOs: African Encounters,* 1982.

——. *UFOs over Africa,* 1997.

Hopkins, Budd. *Intruders,* 1987.

——. *Witnessed: the True Story of the Brooklyn Bridge UFO Abductions,* 1996.

——. "Invisibility and the UFO Abduction Phenomenon." *MUFON Symposium Proceedings,* 1993.

Hopkins, Budd, and Carol Rainey. *Sight Unseen: Science, UFO Invisibility and Transgenic Beings,* 2003.

Howe, Linda. "The UFO Jigsaw." *International Symposium on UFO Research,* 1992.

Huggins, David. "My Close Encounters of the Third Kind." *Leading Edge International Research Group Journal,* Issue No. 46, August–September 1992.

Huneeus, Antonio. "Ufology in the Far East: China, Japan, Korea." *MUFON Symposium Proceeding*, 1997.

Huyghe, Patrick. *The Field Guide to Extraterrestrials*, 1996.

——. *Swamp Gas Times: My Two Decades on the UFO Beat*, 2001.

Hynek, J. Allen. "The 10 Best Cases." *Frontiers of Science*, May–June 1981.

——. *The UFO Experience: A Scientific Inquiry*, 1972.

Jacobs, David, Ph.D. *Secret Life: Firsthand Accounts of UFO Abductions*, 1992.

——. *The Threat*, 1998.

——, ed. *UFOs & Abductions: Challenging the Borders of Knowledge*, 2000.

"Johns, Kate," "The UFO Report: I Was Abducted by Aliens." *New Idea*, November 17, 2001.

Kelleher, Colm. "A Cautionary Tale: DNA Analysis of Alleged Extraterrestrial Biological Material: Anatomy of a Molecular Forensic Investigation," NIDS website: *www.nidsci.org*, 2003.

Kottmeyer, Martin. "Apocalyptic Thought Connected with UFOs and ETs." In *The Encyclopedia of Extraterrestrial Encounters* Ronald Story (ed.), 2001.

Leir, Dr. Roger. *The Aliens and the Scalpel: Scientific Proof of Extraterrestrial Implants in Humans*, 1998.

——. *Casebook: Alien Implants*, 2000.

——. "Alien Abduction & Academic Science," *UFO Magazine*, January 2004.

Leslie, Desmond, and George Adamski. *Flying Saucers Have Landed* 1953, 1970.

Lewis, James, ed. *The Gods Have Landed: New Religions from Other Worlds*, 1995.

Liljegren, Anders. "The Continuing Story of the Imjarvi Skiers," *FSR*, Vol. 26, No. 3 and 5, 1980–81.

Longren, Eve. *The Love Bite: Alien Interference in Human Relationships*, 2000.

Lorenzen, Coral. *Flying Saucers: The Startling Evidence of the Invasion from Outer Space*, 1966.

Lorenzen, Coral and Jim. *Flying Saucer Occupants*, 1967 (reissued in a revised and expanded format as *Encounters with UFO Occupants* in 1976).

Lowrey, Gary. *Alien Abduction: The Gary Lowrey e-book*, undated.

Mack, John. *Passport to the Cosmos: Human Transformation and Alien Encounters,* 1999.

Mandelker, Scott, Ph.D. *From Elsewhere: Being E.T. in America,* 1996.

Matheson, Terry. *Alien Abductions: Creating a Modern Phenomenon,* 1998.

Mebane, Alexander. "The 1957 Saucer Wave in the United States." In Aime Michel, *Flying Saucers and the Straight-Line Mystery,* 1958.

Michel, Aime. "The Strange Case of Dr. 'X,' " (2 parts) in *FSR "UFO Percipients"* Special Issue No. 3, September 1969, and *FSR* Vol. 17, No. 6, 1971.

Montgomery, Ruth. *Aliens Among Us,* 1985.

Musser, Dale. "Anatomy of a Mass Abduction." *UFO Universe,* Fall 1993.

Parry, Simon. "Airline Pilots in Close UFO Encounter over China." *CenturyChina.com,* 25 November 2003.

Pratt, Bob. *UFO Danger Zone: Terror and Death in Brazil: Where Next?* 1996.

Pritchard, Andrea, and David Pritchard, John Mack, Pam Kasey, and Claudia Yapp, eds. *Alien Discussions: Proceedings of the Abduction Study Conference held at MIT,* Cambridge, MA, 1994.

Randle, Kevin, Russ Estes, and William Cone, Ph.D. *The Abduction Enigma: The Truth Behind the Mass Alien Abductions of the Late Twentieth Century,* 1999.

Randle, Kevin, and Russ Estes. *Faces of the Visitors,* 1997.

Randles, Jenny. *The Pennine UFO Mystery,* 1983.

——. *Abduction,* 1988.

——. *The Most UFO-Haunted Places in the World in UFOs and How to See Them,* 1992.

——. *Aliens: the Real Story,* 1993.

——. *Star Children,* 1994.

——. *Investigating the Truth behind MIB: The Men in Black Phenomenon,* 1997.

——. *The Complete Book of Aliens & Abductions,* 1999.

Reader's Digest. *UFO: The Continuing Enigma,* 1991.

Rosario Real, Luiz do. "The New 'A.V.B.' Case from Brazil: Full Account." *FSR* 31/3, 1986, also reported in *FSR* 30/5, 1985.

Ribera, Antonio. "The Soria Abduction," Parts I–III, *FSR,* Vol. 30, No. 3-5, May–June 1984 to Sept.–Oct. 1984.

——. "The Jinn and the Dolmen: The Most Amazing Case of Abduction Yet," *FSR*, Vol 31, No. 4, 1986.

Ring, Kenneth. *The Omega Project: Near Death Experiences, UFO Encounters, and Mind at Large*, 1992.

Rodeghier, Mark. "Monitoring Abductions," *MUFON Journal*, No. 405, January 2002.

Rodwell, Mary. *Awakening: How Extraterrestrial Contact Can Transform Your Life*, 2002.

Romaniuk, Pedro. "Rejuvenation Follows Close Encounter with UFO," *FSR*, Vol. 19, No. 4, July-August 1973.

——. "The Extraordinary Case of Rejuvenation," *FSR*, Vol. 19, No. 5, September-October 1973.

Schnabel, Jim. *Dark White: Aliens, Abductions, and the UFO Obsession*, 1994.

Schuessler, John. *The Cash-Landrum UFO Incident*, 1998.

Sprinkle, Leo, PhD. *Soul Samples*, 1999.

Sherman, Dan. *Above Black: Project Preserve Destiny, Insider Account of Alien Contact and Government Cover-up*, 1997.

Steiger, Brad. *Gods of Aquarius: UFOs and the Transformation of Man*, 1976.

——. *The Star People*, 1981.

——. *The Seed*, 1983.

Strieber, Whitley. *Communion: A True Story*, 1987.

——. *Transformation*, 1988.

——. *Breakthrough*, 1995.

——. *The Secret School: Preparation for Contact*, 1997.

——. *Confirmation the Hard Evidence of Aliens Among Us*, 1998.

Strieber, Whitley, and Anne Strieber, *The Communion Letters*, 1997.

Sturrock, Peter, ed. *The UFO Enigma: A New Review of the Physical Evidence*, 1999.

Swords, Michael. "Extraterrestrial Hybridization Unlikely." *MUFON Journal*, November, 1988, and "Modern Biology and the Extraterrestrial Hypothesis" *MUFON 1991 International UFO Symposium proceedings*, abbreviated in "Modern Biology and Extraterrestrials," in Bill Fawcett's *Making Contact*, 1997.

Taylor, Barry. "DNA Testing Inconclusive: Mystery Prints on Mirror Investigated," *MUFON UFO Journal*, November 2000.

Thompson, Richard. *Alien Identities: Ancient Insights into Modern UFO Phenomena,* 1993.

Turnage, C. L. *Sexual Encounters with Extraterrestrials,* 2001.

Vallee, Jacques. *Passport to Magonia: From Folklore to Flying Saucers,* 1969.

———. *Confrontations: A Scientist's Search for Alien Contact,* 1990.

———. *Forbidden Science: Journals (1957–69),* 1996.

Walton, Travis. *Fire in the Sky: the Walton Experience,* 1996.

Wilson, Colin. *Alien Dawn: An Investigation into the Contact Experience,* 1998.

Wisner, Kerry, and Linda Rakos. *Reveal Nothing: A Forensic Investigation of a UFO Encounter,* 2001.

Other

Abanes, Richard. *End-Time Visions: the Road to Armageddon?* 1998.

Andrews, Colin, with Stephen Spignesi. *Crop Circles: Signs of Contact,* 2003.

Atwater, F. Holmes. *Captain of My Ship, Master of My Soul,* 2001.

Benjamin, Marina. *Living at the End of the World,* 1999.

Bord, Janet. *Fairies: Real Encounters with Little People,* 1997.

Briggs, Katharine. *A Dictionary of Fairies,* 1977.

Brown, Courtney. *Cosmic Voyage: True Evidence of Extraterrestrials Visiting Earth,* 1996.

———. *Cosmic Explorers: Scientific Remote Viewing, Extraterrestrials, and a Message for Mankind,* 2000.

Clancy, Susan, and Richard McNally, Daniel Schacter, Mark Lenzenweger, and Roger Pitman. "Memory Distortion in People Reporting Abduction by Aliens." *Journal of Abnormal Psychology,* Vol. 111, No. 3, 2002.

Clark, Jerome. "Fairies." In *Unexplained! 347 Strange Sightings, Incredible Occurrences, and Puzzling Physical Phenomena,* 1993.

Clarke, David. *Supernatural Peak District,* 2000.

Clarke, David, with Andy Roberts, *Twilight of the Celtic Gods: An Exploration of Britain's Hidden Pagan Traditions,* 1996.

Cooke, Jennifer. *Cannibals, Cows and the CJD Catastrophe,* 1998.

Crawford, I. M. *The Art of the Wandjina,* 1968.

Devereux, Paul, John Steele, and David Kubrin. *Earthmind,* 1989.

Devereux, Paul. *Places of Power,* 1990.

Doore, Gary, ed. *Shaman's Path: Healing, Personal Growth and Empowerment,* 1988.

Drury, Nevill. *Shamanism,* 1996.

——. *Vision Quest: A Personal Journey through Magic and Shamanism,* 1984.

Crowley, Brian, and James Hurtak. *The Face on Mars,* 1986.

Crowley, Brian, and Anthony Pollack. *Return to Mars,* 1989.

Curran, Douglas. *In Advance of the Landing: Folk Concepts of Outer Space,* 1985.

Evans, Christopher. *Cults of Unreason,* 1973.

Everson, Michael. "Tenacity in Religion, Myth, and Folklore: The Neolithic Goddess of Old Europe Preserved in a Non-Indo-European Setting." *Journal of Indo-European Studies* 17, Nos. 3 and 4, Fall/Winter 1989, pgs. 277–95.

Freeman, Richard. "In the Coils of the Naga." *Fortean Times* 166, January 2003.

Gregory, Lady. *Irish Myths and Legends* (various editions, which incorporate *Gods and Fighting Men: The Story of the Tuatha de Danaan and of the Fianna of Ireland*), 1998.

Grey, George. *Expeditions in Western Australia 1837–1839,* 1983.

Grof, Christina, and Stanislav Grof. *The Stormy Search for the Self: Understanding and Living with Spiritual Emergency,* 1991.

Halpern, Paul. *Countdown to Apocalypse,* 1998.

Harner, Michael. *The Way of the Shaman,* 1980.

Holland, Heidi. *African Magic,* 2001.

Hough, Peter, and Jenny Randles. *Looking for the Aliens,* 1991.

Howe, Linda Moulton. *Mysterious Lights and Crop Circles,* 2000.

Hufford, David. *The Terror that Comes in the Night: An Experience-Centred Study of Supernatural Assault Traditions,* 1982.

Jian, Ma. *Red Dust: A Path through China,* 2002.

Kalweit, Holger. *Dreamtime and Inner Space: The World of the Shaman,* 1988.

Kaminer, Wendy. *Sleeping with Extra-terrestrials: The Rise of Irrationalism and Perils of Piety,* 1999.

Keeney, Bradford. *Shaking Out the Spirits,* 1994.

Keeney, Bradford, ed. *Vusamazulu Credo Mutwa: Zulu High Sanusi,* 2001.

Kurlansky, Mark. *The Basque History of the World,* 2000.

Larsen, Stephen, ed. *Song of the Stars: The Lore of a Zulu Shaman,* 1996.

Lawlor, Robert. *Voices of the First Day: Awakening in the Aboriginal Dream-time,* 1991.

Lieb, Michael. *Children of Ezekiel: Aliens, UFOs, the Crisis of Race, and the Advent of End Time,* 1998.

Lifton, Robert Jay. *Destroying the World to Save It: Aum Shinrikyo, Apocalyptic Violence, and the New Global Terrorism,* 1999.

Luna, Luis Eduardo, and Pablo Amaringo. *Ayahuasca Visions: The Religious Iconography of a Peruvian Shaman,* 1991.

Mackillop, James. *Oxford Dictionary of Celtic Mythology,* 1998.

McKenna, Terence. *True Hallucinations: Being an Account of the Author's Extraordinary Adventures in the Devil's Paradise,* 1993.

Miller, Russell. *Bare Faced Messiah: the True Story of L. Ron Hubbard,* 1987.

Moody, Raymond, with Paul Terry. *Reunions: Visionary Encounters with Departed Loved Ones,* 1994.

Mowaljarlai, David, and Dutta Malnic. *Yorro Yorro: Everything Standing Up Alive—Spirit of the Kimberley,* 1993.

Muller, Klaus, and Ute Ritz-Muller. *Soul of Africa: Magical Rites and Traditions,* 1999.

Narby, Jeremy. *The Cosmic Serpent: DNA and the Origins of Knowledge,* 1999.

Narby, Jeremy, and Francis Huxley, eds. *Shamans Through Time: 500 Years on the Path to Knowledge,* 2001.

O'Neill, John. "Nocturnal Visitors." *The Skeptic,* Spring 2000, Vol. 20, No. 3, pgs. 30–33.

Pendergrast, Mark, *Mirror|Mirror: A History of the Human Love Affair with Reflection,* 2003.

Purkiss, Diane. *Troublesome Things: A History of Fairies and Fairy Stories,* 2000.

Randles, Jenny. *Time Storms: Amazing Evidence for Time Warps, Space Rifts and Time Travel,* 2001.

——. *Supernatural Pennines,* 2002.

Ronson, Jon. *Them: Adventures with Extremists,* 2001.

(Russell, George), Æ. *The Candle of Vision,* 1965.

Rux, Bruce. *Architects of the Underworld,* 1996.

Schechter, Danny. *Falun Gong's Challenge to China: Spiritual Practice or "Evil Cult"? A Report & Reader,* 2001.

Schnabel, Jim. *Remote Viewers: The Secret History of America's Psychic Spies,* 1997.

Showalter, Elaine. *Hystories: Hysterical Epidemics and Modern Culture,* 1997.

Siegel, Ronald. *Fire in the Brain,* 1992, 1993.

Stewart, R.J. (compiler). *Robert Kirk: Walker Between Worlds* (a new edition of *The Secret Commonwealth of Elves, Fauns & Fairies*), 1990.

Stone, Jon, ed. *Expecting Armageddon; Essential Readings in Failed Prophecy,* 2000.

Swann, Ingo. *Penetration: The Question of Extraterrestrial and Human Telepathy,* 1998.

Targ, Russell, and Harold Puthoff. *Mind-Reach: Scientists Look at Psychic Abilities,* 1977.

Temple, Robert. *The Genius of China: 3,000 Years of Science, Discovery and Invention,* 2002.

Tyler, Christian. *Wild West China: The Taming of Xinjiang,* 2003.

Von Denmarc, Jan. *Wisdom from the Great Beyond: A True Story of Life with a Ghost,* 2003.

Weber, Eugen. *Apocalypses: Prophecies, Cults & Millennial Beliefs Through the Ages,* 1999.

Winton, Tim. *That Eye, the Sky,* 1986.

Wolf, Fred Alan. *The Eagle's Quest: A Physicist's Search for Truth in the Heart of the Shamanic World,* 1991.

Wood, Frances. *The Silk Road: Two Thousand Years in the Heart of Asia,* 2003.

INDEX

If you believe you have legitimate biological evidence of a UFO abduction experience, you are encouraged to contact the Anomaly Physical Evidence Group through Bill Chalker. Any evidence presented to us will be assessed for its potential as credible evidence warranting the cost, resources, and time involved in its study. The APEG was formed to focus attention on biological strategies in abduction investigations. Preliminary funding has enabled a small laboratory presence addressing this exciting area of research. The APEG can be contacted through this writer at P.O. Box 42, West Pennant Hills, NSW, 2125, Australia. Further funding of this research is encouraged. Please contact Bill Chalker for details.